Atoms and Molecules in Laser and External Fields

Atoms and Molecules in Laser and External Fields

Editor
Man Mohan

Alpha Science International Ltd.
Oxford, U.K.

Atoms and Molecules in Laser and External Fields
186 pgs. | 5 tbs. | 94 figs.

Editor
Man Mohan
Department of Physics and Astrophysics
University of Delhi
Delhi, India

Copyright © 2008

ALPHA SCIENCE INTERNATIONAL LTD.

7200 The Quorum, Oxford Business Park North
Garsington Road, Oxford OX4 2JZ, U.K.

www.alphasci.com

Printed from the camera-ready copy provided by the Author.

ISBN 978-1-84265-513-9

Printed in India

Preface

The recent development of highly advanced lasers and other, external field sources (e.g., Synchrotron radiation, strong electric and magnetic fields)are opening up new frontier of science which cover explorations in broad area ranging from atomic molecular and optical physics to nuclear physics and high energy particle physics; even further to astrophysics and cosmology. The importance of Atomic, Molecular an Optical Physics is well evident from the list of Nobel Prize winners in the past couple of years. The Noble Prize for Physics 2005 was awarded jointly to R.J. Glauber, J.L. Hall and T.W. Hansch for pioneering theoretical and experimental work on Laser Physics including optical coherence and laser-based precision spectroscopy. Laser cooling and Bose-Einstein condensation are some of the hot topics these days. W. Ketterle, E.A. Cornell land C.A. Wieman were awarded the Nobel prize in 2001 for their discovery of the new state of matter-Bose-Einstein Condensation. In 1997, S. Chu, C. Cohen-Tannoudji and W.D. Phillips were awarded the Nobel Prize for their developing methods to cool and trap atoms with laser light. Further, for the studies of the transition states of chemical reactions using femtosecond spectroscopy, the Nobel Prize was awarded in 1999, to A.H. Zewial. In the optical and near infrared frequency range ($\hbar\omega = 1$ e.V), the next generation of high power lasers will reach the intensity level of 10^{23} W/c^2. At VUV frequencies ($\hbar\omega = 100$ e.V.), a record level of almost 10^{16} w/cm^2 has recently been achieved at the FLASH facility (DESY, Germany) with a free-electron laser. In the near future efforts will be made to produce ultrashort atto-second (1 a.s. $= 10^{-18}$ sec.), high frequency radiation ($\hbar\omega = 10$ to 1000 e.V.) from plasma surface harmonics where, due to high conversion efficiency, considerable higher intensities might be reachable. With such type of light sources one no longer merely observes nature, but can reshape and redirect atoms,molecules, particles or radiation. This new drive towards harnessing quantum dynamics is enormously important to future developments in fundamental physics and applied energy science. Few of the several topics covered in the book are; Initiating strong field processes using Atto-second pulses, Trapping Population inside a nano-structure potential well, Higher Harmonic Generation (HHG), Molecular Dynamics under control with photonic Reagents etc.

This book will be useful for graduate students, Researchers, Scientists and Teachers in the Universities, IIT's, National and International Laboratories.

I am thankful to Prof. Nicholas Blombergen, Noble Laurette and Prof. H. Walther for giving the opportunity for working in the University of Munich and Max Plnck Institute and for sending the best wishes for the success of the mission. I am also indebted to Prof. Sreenivasan (Director, ICTP, Trieste, Italy), Prof. S.K. Joshi, (Ex-Director, N.P.L.), for their encouragement. I am also thankful to our Vice Chancellor Prof. Deepak Pental (Pro-Vice-Chancellor, Delhi University), Prof. S.K. Tandon, Dr. B.S. Singh (Principal, K.M. College, Delhi University), Mr. I.M. Kaphai and my all colleagues of K.M. College for constant help and encouragement. I am also thankful to my colleagues at Delhi University, Prof. D.S. Kulshreshta, Prof. R.P. Tandon, Prof. K.L. Baluja and research students Dr. Vinod Prasad, Dr. Avnindra Kumar, Dr. Pardeep K. Jha, Dr. Nupur Verma, Dr. Alok K.S. Jha and Dr. V. Ranjan for helping me in making the book. I am also thankful to Mr. Gaurav and Mr. Saurav Chaddha of Bros Communication Pvt. Ltd, and M/s Scientific Documentations for shaping the articles in the style of the book. I am thankful to wife Mrs. Sneh Mohan, daughter Anu and son Anurag of their constant encouragement.

The Editor

Contents

Atoms and Molecules in Laser and External Fields
Editor: Man Mohan
Copyright © 2008, Narosa Publishing House, New Delhi, India

Placing Molecular Dynamics under Control with Photonic Reagents

Herschel Rabitz

Department of Chemistry, Princeton University, Princeton 08544, U.S.A.

BACKGROUND

Molecular dynamics encompasses many phenomena and applications, but the heart of the subject lies in making chemical transformations. There is also much interest in creating special dynamical states of matter, even if they only have a fleeting existence. This paper concerns the emerging role of radiation in the form of shaped laser pulses behaving as a new class of "photonic reagents" capable of altering atomic- and molecular-scale events[1]. Radiation at specific wavelengths can be used to influence atomic and molecular scale dynamics[2]. However, the freedom available in adjusting the laser intensity and carrier frequency generally is too limiting to provide selective manipulation, especially in cases where many product channels are accessible. A shift in thinking on the potential value of lasers for manipulating atomic and molecular scale events has occurred in recent years. Radiation interacts with atoms and molecules primarily through its electric field[3] acting on the electrons and nuclei. This radiation-matter coupling is analogous to Coulombic driven atom-atom interactions, which form the basis of traditional chemical reagent operations.

The initial thinking some forty years ago, with regard to the possible enhanced role of radiation in facilitating molecular-scale transformations, viewed a laser beam as having the two basic characteristics of being monochromatic and intense. The dream for laser-driven selective molecularscale transformations and excitations had enormous appeal due to its apparent simplicity. Experimentation along these lines started around 1970 in both academic and industrial laboratories. Roughly, a decade of study followed with much frustration being the main outcome. First, no doubt existed that appropriate excitation was taking place, as this process rests on the fundamental tenet of spectroscopy that a local molecular feature (e.g., a bond in a molecule) has a unique spectral signature. However, a bond is not isolated from the other surrounding atoms in the molecule or the solvent. Thus, although radiation may be entering the molecule at the right place, the deposited energy may disperse at a high rate to activate other bonds. These apparently uncontrolled intra-molecular energy transfer processes seemed to present a vexing problem without obvious solution. The shaping of laser pulses to act as photonic reagents has provided a means to manage the dynamical processes involved as explained below.

LASER CONTROLLED MOLECULAR-SCALE PROCESSES

The modern era of laser control over atomic and molecular processes may be traced to the development of new concepts and tools introduced in the 1980's and 1990's. First came the recognition that manipulating complex molecules with radiation should best be expressed as a problem in (quantum) control engineering. Using engineering principles, one would naturally be led to consider (a) performing a computational design of the particular laser pulse shape for making the specified molecular-scale transformation, and then (b) constructing the shaped laser pulse in the laboratory to act as a photonic reagent hopefully for its successful application to a sample of the molecules. Except for some very simple proof-of-principle experiments, the subject did not evolve along this "design and construct" paradigm. The difficulty that blocked this route may be easily understood. First, the fundamental basis of quantum control rests on manipulating wave functions to steer the dynamics towards one product versus that of another product (e.g., break one bond in competition with another bond in a molecule). Computational design of a laser pulse shape for this delicate manipulation generally depends sensitively on having a quantitative knowledge of the interaction of the radiation with the molecule as well as a full understanding of the atomic interactions within the molecule. For virtually all molecules of interest, there is considerable quantitative uncertainty about the interactions involved. Furthermore, accurately solving the design equations (i.e., the time-dependent Schrdinger equation) is very difficult for polyatomic molecules, especially with intense fields. Although the design of successful photonic reagents generally remains out of reach at the present time, considerable physical insight has been learned from carrying out such computational design studies[4,5] starting in the late 1980's. Efforts along these lines continue with much still being learned about the principles of laser control of matter[6].

Although the reliable a priori design of laser pulse shapes for manipulating molecules is generally blocked, the means to proceed in the laboratory towards successful control resides in the laser tools. Successful molecular manipulation is generally expected to require a laser pulse (t) depending on time t containing many frequency components, where each has just the right coordinated amplitude and phase, to push and pull on the atoms and electrons in sync with their natural motions. In this fashion, the specially shaped laser pulse can cooperate, and when necessary, compete with the natural atomic scale forces to gently, or even violently, steer the dynamics towards the right product. These rather complex characteristics of successful molecular control laser fields were also consistently found in the 5 theoretical design calculations.

A shaped pulse may be synthesized by first starting with an unshaped ultrafast laser pulse obtained directly from the laser[7,8]. The unshaped raw pulse is typically merely femtoseconds in duration and usually featureless with overall Gaussian shape. The next task is to adjust the phase and amplitude of the many frequency components inherently present in the raw pulse to sculpt a particular pulse shape tailored to act as a photonic reagent.

A crucial element in finding effective photonic reagents lies in the high duty cycle of the laser and pulse shaper, with their capability of going from one trial pulse shape to another at the rate of even hundreds or more per second[7]. A single experiment would consist of creating a particular trial laser pulse shape, applying it to a molecule or material of interest, followed by the detection (perhaps with a second laser pulse) of the outcome from the controlled molecular-scale dynamical event. Such a single experiment might occur in a second or less, including signal averaging, at which point the apparatus would be primed for a follow-on experiment with a new trial laser pulse shape, etc. This high duty cycle feature can be exploited by suitable pattern recognition software capable of observing the prior experiment(s) and rapidly suggesting a new, and hopefully, improved one for execution. Many types of pattern recognition software exist, but genetic or related evolutionary algorithms have been[9] the choice thus far. High-speed pattern recognition software is necessary to rapidly make the decisions on how to adjust the laser after each experiment. Experiments of this type were first suggested[10] in 1992, followed by several years of simulations indicating their potential capabilities until the first actual laboratory study[11] was undertaken in 1997. The closed-loop operation of the experiment allows the molecules to iteratively "teach the laser" exactly which pulse shape is just right for their desired manipulation[10].

THE CAPABILITIES OF SHAPED LASER PULSES ACTING AS PHOTONIC REAGENTS

A basic issue for assessment is the capability of shaped laser pulses to act as fleeting photonic reagents for creating specialized excitations or even permanently altering a molecular material[1]. To appreciate the potential utility of photonic reagents, an important consideration is that a typical control laser pulse can have an intensity $\sim 10^{14}$ W/cm^2 or even more. Although the laser pulse might last for only $\sim 10^{-13}$ seconds, during that time, the atoms and electrons will feel enormous forces that are fully competitive with naturally occurring atomic-scale forces. Therein lies the basic capability of manipulating molecular-scale events with intense laser pulses, where their shaping manages the forces to carefully steer about molecular motion in a desired fashion.

ATOMS AND MOLECULES IN LASER AND EXTERNAL FIELDS

All of the current experiments employ the Ti: Sapphire laser as the source of the pulses for shaping. The Ti: Sapphire laser operates at a central wavelength ~ 800 nm and a bandwidth of ~ 20 nm due to its pulsed nature[13]. However, the available laser bandwidth alone would typically not be sufficient to perform significant high quality selective manipulation of dynamical events. To overcome this difficulty, many current experiments operate using an intense pulse, which severely disturbs the molecular motion and deposits many photons in a hopefully beneficial manner[14]. The high intensity of the pulse will cause dynamic power broadening of the molecular energy levels to take them in or out of resonance with the radiation, as best suits the goal of reaching the specified dynamical objectives. In this fashion, effective dynamical bandwidth can be created and exploited. The many emerging successful experiments[11,12,14–22] attest to the capabilities of this technique, especially by playing on the large numbers of laser pulse shaper pixels to tailor the photonic reagents to have the right structure to meet the posed molecular-scale objectives. The controlled dynamical events demonstrated thus far range from manipulating the electronic structure of atoms out through altering energy transfer in complex biomolecules. Below is a summary of a few of the emerging experiments exploiting these tools.

In 1997, Wilson and coworkers[11] first demonstrated the closed-loop control operations to identify the nature of the laser pulses needed to maximize the fluorescence signal from a laser dye. The experiment was set up with a known answer for a rather simple system to test the control concept. This basic experiment was seminal for all that has followed in the laboratory closed-loop quantum control applications. It is also now known that many distinct pulse shapes can produce the same final product state in most applications[23]. This prospect has important practical and fundamental implications. The breaking of one chemical bond versus that of another with lasers was one of the principal challenges presented some 40 years ago. Initial successful experiments of this type were carried out[15] in 1998 on the molecule $CpFe(CO)_2Cl$. Since then many additional illustrations of selective bond breaking have been demonstrated. The experiments are typically initiated with a family of random trial shaped laser pulses, and the output signal for optimization may be a mass spectrum or other signature of the fragmentation. Besides breaking one bond versus another, an interesting and more complex goal is the selective dissociative rearrangement of polyatomic molecules. Such an experiment was carried out on acetophenone[14,24], whereby it was converted to toluene. This process entails breaking the phenyl-carbon and methyl-carbon bonds, as well as forming the methyl-phenyl bond all simultaneously.

The field of chemistry operates with a remarkably small set of reaction classes producing an enormous outcome of distinct products. A fundamental question is whether this notion carries forth to photonic reagents permitting to view them as having their own systematic "chemistry". An intriguing experiment in this regard[19] involved the removal of CO from the organomettalic compound $CpFe(CO)_2X$ that was halogenated with either X=Cl or Br. A shaped photonic reagent was found for optimizing the removal of CO from the chlorinated compound, and it was then applied to the brominated compound, where it worked successfully but not fully optimally. In turn, the shaped photonic reagent determined to work optimally on the brominated compound also successfully worked to remove CO from the chlorinated compound, but again, not optimally. This collective behavior is exactly what one would hope to find when thinking of shaped laser pulses as photonic reagents with systematic properties. The generality of this concept needs thorough exploration with broad varieties of molecular-scale goals, while working with various homologous molecules.

In the illustrations above, the control product was some type of molecular transformation. In the present case, the control outcome is radiation at a high harmonic of the input laser pulse central frequency ω. As background to the challenge involved, it had been known for many years that an intense laser pulse passing through a gas can generate[25] typically, many hundreds of high harmonics of ω. A longstanding goal in this area had been to direct the output radiation into a single high harmonic while simultaneously not inducing additional excitation of other harmonics. An experiment of this. Atoms and Molecules[2] in Laser and External Fields type was successfully performed[16] in argon, optimizing the signal in the 27th harmonic corresponding to emitted soft X-ray radiation. In this case, a detailed model of the process was available and a consistent physical picture of the mechanism was attained. Certain special features of the shaped driving pulse were identified as critical for focusing into the 27th high harmonic. Mechanistic studies of this type[26] are just beginning to appear in other applications of photonic reagent control. Just as in molecular transformations with traditional reagents, the mechanisms by which shaped photonic reagents achieve success is of keen interest.

Finally, one general issue is the ability of photonic reagents to operate effectively with highly complex molecules. One such example concerns light-induced electron transfer in the antenna complex of photosynthetic bacteria. The application of traditional unshaped laser pulses was largely unsuccessful in changing the electron transfer process. However, application of the closed-loop control concept[22] was able to affect such a change. Various optical probes were used as feedback signals to guide the shaping of the laser pulse for redirecting the excitation pathway through the antenna complex. In the experiment, the goal was to enhance the internal energy transfer conversion process over that of the electron transfer process. This ratio was manipulated by some 30%, and a detailed examination of the laser pulse structure revealed much information about key aspects of the antenna complex excitation mechanism. This is only an initial step towards employing closed-loop control to both manipulate and understand biological systems.

WHY AND HOW DO PHOTONIC REAGENTS OPERATE?

The closed-loop laser machine architecture provides the means to control quantum phenomena with many emerging successes, but these outcomes in themselves do not explain why the experiments work or how the quantum dynamics evolve under control. Addressing these two questions of why and how the experiments operate is fundamental to projecting the likely future directions of the field.

Why the experiments actually succeed, and in a relatively easy fashion, appears to be puzzling upon considering a few facts. First, in the laboratory typically hundreds of phase and amplitude control variables with grey scale settings are searched over to find an effective photonic reagent. The simple counting of digitized settings of these variables reveals that an astronomical number of experiments could potentially be performed. Yet, in virtually all cases, merely hundreds to thousands of experiments carried out on the order of minutes typically suffice to significantly enhance the desired product yield. This behavior defies what is commonly called the "curse of dimensionality". Second, many experiments involve high order multi-photon processes, which can be sensitive to the intensity of the shaped and focused laser pulse as well as the orientation of the molecules relative to the laser polarization. Such pulses also have spatial profiles both parallel and transverse to their propagation direction implying that the intensity of the photonic reagents will be spatially inhomogeneous. Intuition would suggest that molecules only in certain locations and with favorable orientations would have just the right intensity to properly effect the control process. Fighting against both the large number of control variables and spatial effects would seem to imply that the control experiments would be very difficult to perform successfully. A full explanation of the observed attractive behavior (i.e., the experiments actually work and with relative ease) calls for further research, but the basic reasons are beginning to emerge. First, fundamental analysis of the quantum control landscape (i.e., physical 23 objective as a function of the control variables) reveals that there are no false traps to be caught in. This simple landscape structure arises in the clean case of attempting to maximize the probability of going from one quantum state to another. In this case, the control landscape of the transition probability only has extrema taking on the values of zero (i.e., no control) and one (i.e., perfect control). Qualitatively, this landscape analysis provides a basis to explain how the curse of dimensionality is beat when performing quantum control experiments (i.e., there are no false traps to fall into). The landscape analysis in itself does not reveal why control spatial inhomogeneities can be 2. Atoms and Molecules in Laser and External Fields 16 type was successfully performed in argon, optimizing the signal in the 27th harmonic corresponding to emitted soft X-ray radiation. In this case, a detailed model of the process was available and a consistent physical picture of the mechanism was attained. Certain special features of the shaped driving pulse were identified as critical for focusing into the 27th high harmonic. Mechanistic studies of this type[26] are just beginning to appear in other applications of photonic reagent control. Just as in molecular transformations with traditional reagents, the mechanisms by which shaped photonic reagents achieve success is of keen interest[27].

Finally, one general issue is the ability of photonic reagents to operate effectively with highly complex molecules. One such example concerns light-induced electron transfer in the antenna complex of photosynthetic bacteria. The application of traditional unshaped laser pulses was largely unsuccessful in changing the electron transfer process. However, application of the closed-loop control concept[22] was able to affect such a change. Various optical probes were used as feedback signals to guide the shaping of the laser pulse for redirecting the excitation pathway through the antenna complex. In the experiment, the goal was to enhance the internal energy transfer conversion process over that of the electron transfer process. This ratio was manipulated by some 30%, and a detailed examination of the laser pulse structure revealed much information about key aspects of the antenna complex excitation mechanism. This is only an initial step towards employing closed-loop control to both manipulate and understand biological systems.

WHY AND HOW DO PHOTONIC REAGENTS OPERATE?

The closed-loop laser machine architecture provides the means to control quantum phenomena with many emerging successes, but these outcomes in themselves do not explain why the experiments work or how the quantum dynamics evolve under control. Addressing these two questions of why and how the experiments operate is fundamental to projecting the likely future directions of the field. Why the experiments actually succeed, and in a relatively easy fashion, appears to be puzzling upon considering a few facts. First, in the laboratory typically hundreds of phase and amplitude control variables with grey scale settings are searched over to find an effective photonic reagent. The simple counting of digitized settings of these variables reveals that an astronomical number of experiments could potentially be performed. Yet, in virtually all cases, merely hundreds to thousands of experiments carried out on the order of minutes typically suffice to significantly enhance the desired product yield. This behavior defies what is commonly called the "curse of dimensionality". Second, many experiments involve high order multi-photon processes, which can be sensitive to the intensity of the shaped and focused laser pulse as well as the orientation of the molecules relative to the laser polarization. Such pulses also have spatial profiles both parallel and transverse to their propagation direction implying that the intensity of the photonic reagents will be spatially inhomogeneous. Intuition would suggest that molecules only in certain locations and with favorable orientations would have just the right intensity to properly effect the control process. Fighting against both the large number of control variables and spatial effects would seem to imply that the control experiments would be very difficult to perform successfully. A full explanation of the observed attractive behavior (i.e., the experiments actually work and with relative ease) calls for further research, but the basic reasons are beginning to emerge. First, fundamental analysis of the quantum control landscape (i.e., physical objective as a function of the control variables)[23] reveals that there are no false traps to be caught in. This simple landscape structure arises in the clean case of attempting to maximize the probability of going from one quantum state to another. In this case, the control landscape of the transition probability only has extrema taking on the values of zero (i.e., no control) and one (i.e., perfect control). Qualitatively, this landscape analysis provides a basis to explain how the curse of dimensionality is beat when performing quantum control experiments (i.e., there are no false traps to fall into). The landscape analysis in itself does not reveal why control spatial inhomogeneities can be 2. Atoms and Molecules in Laser and External Fields overcome. This feature appears to be due to the existence of innumerable control solutions for meeting any particular target objective. The fact that multiple, and typically an infinite number, of control solutions exist for meeting a posed objective, opens up the opportunity to search amongst these controls in the laboratory for at least one that simultaneously works successfully regardless of the spatial location of the target molecule. A theoretical analysis of this situation reveals that such a nexus of control solutions generally exists, such that the optimal control experiments work to sweep all of the 28 molecules to a common target regardless of where they are located in the laser beam spatial profile.

Understanding why the experiments work, does not explain how they operate (i.e., quantum control mechanisms). As mentioned earlier, mechanistic insights are beginning to become available to explain the physical processes operating[16,26,27] during controlled quantum dynamics. There are many challenges in this area, including merely defining mechanism in the context of quantum coherence being a central factor. In addition, the existence of multiple control solutions capable of meeting the same final physical objective[23] must be kept in mind. Although a given control experiment will end up discovering one of these controls and an associated mechanism for its action will exist, many other controls producing distinct mechanisms will likely be equally effective. Thus, it is generally expected that quantum controls and mechanistic interpretations will both not be unique. Rather than being a difficulty, this prospect should be viewed positively with the option now open to search amongst the possible controls for those producing desirable mechanistic, robustness, or other system dynamical properties.

The performance of the control experiments does not have to wait for a complete understanding of why and how they operate, as evident from the growing number of successful control studies. However, achieving mechanistic understanding is fundamental, and the explanations are beginning to fall into place.

WHAT MAY LIE AHEAD FOR UTILIZING PHOTONIC REAGENTS?

Now that shaped photonic laser control is on firm theoretical and implementational footing, the subject is open for full development in the chemical and physical sciences. The core of this development centers around thinking in terms of shaped laser pulses as photonic reagents. Regardless of particular applications, a fundamental task ahead is to establish the rules and understanding of how photonic reagents operate under a variety of circumstances. The current successful experiments hold out great promise that photonic reagents can become flexible tools. Beyond the controlled alteration of molecules or materials with photonic reagents, there remains the general desire to understand how these processes occur (i.e., the mechanisms involved). Exploring mechanisms is likely to benefit from the power of performing massive numbers of experiments in short periods of laboratory time. Finally, there is the opportunity to turn these photonic reagent tools around as a means to better understand molecules and materials. The opportunity here lies in the fact that an ability to control, in principle, provides the means to understand. There is also much to be learned about the fundamental reasons why the experiments can readily find effective control solutions, despite searching over hundreds of pulse shape control variables.

In summary, although the suggestion of employing lasers as molecular-scale manipulation tools goes back some 40 years, in many respects the subject should be viewed as only a few years young. Much of this 40-year period was occupied with establishing the conceptual foundations of the field and bringing online the appropriate technologies. Only in the last few years has all of this come together with significant laboratory demonstrations. Thus, the field is truly young and open for exploration and development.

REFERENCES

1. H. Rabitz, Science, 299, 525 (2003).
2. N. Turro, Modern Molecular Photochemistry, University Science Books, 1991.
3. O. Jefimenko, Electricity and Magnetism: An Introduction to the Theory of Electric and Magnetic Fields (Electret Scientific Co., 2nd Ed., Star City 1989).
4. R. Rice and M. Zhao, Optical Control of Molecular Dynamics, John Wiley and Sons, New York, 2000.
5. A. Peirce, M. Dahleh and H. Rabitz, Phys. Rev. A, 37, 4950 (1988); S. Shi, A. Woody and H. Rabitz, J. Chem. Phys., 88, 6870 (1988); R. Kosloff, S. Rice, P. Gaspard, S. Tersigni and D. Tanner, Chem. Phys., 139, 201 (1989).
6. E. Brown and H. Rabitz, J. Math. Chem., 31, 17 (2002).
7. C. Froehly, U. Colombea and M. Vampouilee, Prog. Opt., 20, 65 (1983); A. Weiner, Prog. Quantum Electron., 19, 161 (1995); A. Weiner, Rev. Sci. Instrum., 71, 1929 (2000).
8. J. Tull, M. Dugan and W. Warren, Adv. Magn. Opt. Reson., 20, 1 (1997).
9. D. Goldberg, Genetic Algorithms in Search, Optimization and Machine Learning, Addison-Wesley, Reading, MA, 1989.
10. R. Judson and H. Rabitz, Phys. Rev. Lett., 68, 1500 (1992).
11. C. Bardeen, V. Yakovlev, K. Wilson, S. Carpenter, P. Weber and W. Warren, Chem. Phys. Lett., 280, 151 (1997).
12. T. Brixner, N. Damrauer and G. Gerber, Advances in Atomic, Molec., Opt. Phys., 46, 1 (2001).
13. D. Strickland and G. Mourou, Optics Comm, 56, 219 (1985).
14. R. Levis, G. Menkir and H. Rabitz, Science, 292, 709 (2001).
15. A. Assion, T. Baumert, M. Bergt, T. Brixner, B. Kiefer, V. Seyfried, M. Strehle and G. Gerber, Science, 282, 919 (1998).
16. R. Bartels, S. Backus, E. Zeek, L. Misoguti, G. Vdovin, I. P. Christov, M. Murnane and H. Kapteyn, Nature, 406, 164 (2000).
17. S. Vajda, A. Bartelt, E. Kaposta, T. Leisner, C. Lupulescu, S. Minemoto, P. Rosendo-Francisco and L. Woste, L. Chem. Phys., 267, 231 (2001).
18. T. Weinacht, J. Ahn and P. Bucksbaum, Nature, 397, 233 (1999).
19. M. Bergt, T. Brixner, C. Dietl, B. Kiefer and G. Gerber, J. Organometallic Chem., 661, 199 (2002).
20. T. Weinacht, J. White, P. Bucksbaum, J. Phys. Chem. A., 103, 10166 (1996b).
21. J. Kunde, B. Baumann, S. Arlt, F. Morier-Genoud, U. Siegner and U. Keller, Appl. Phys. Lett.,77, 924 (2000).
22. J. Herek, W. Wohlleben, R. Cogdell, D. Zeidler and M. Motzkus, Nature, 417, 533 (2002).
23. H. Rabitz, M. Hsieh and C. Rosenthal, Science, 303, 998 (2004).

24. R. Levis and H. Rabitz, J. Phys. Chem. A, 106, 6427 (2002).
25. M. Lewenstein, P. Balcou, M. Ivanov and P. Corkum, Phys. Rev. A, 49, 2117 (1993).
26. C. Daniel, J. Full, L. Gonzalez, C. Lupulescu, J. Manz, A. Merli, S. Vajda and L. Woste, Science, 299, 536 (2003).
27. A. Mitra and H. Rabitz, Phys. Rev. A, 67, 033407 (2003).
28. H. Rabitz and G. Turinici, to be published.

Atoms and Molecules in Laser and External Fields
Editor: Man Mohan
Copyright © 2008, Narosa Publishing House, New Delhi, India

Experimental Investigation of the Optical Response of Nanostructured Metal Surfaces

J. Weiner

Université Paul Sabatier IRSAMC/LCAR 118 route de Narbonne 31062 Toulouse, France

INTRODUCTION

Initial reports of dramatically enhanced transmission through arrays of subwavelength holes in thin films and membranes1 have focused attention on the physics underlying this surprising optical response. Since the early experiments were carried out on metal films, surface plasmon polaritons were invoked to explain the anomalously high transmission and to suggest new types of photonic devices. Other interpretations based on "dynamical diffraction" in periodic slit and hole arrays[2] or various kinds of resonant cavity modes in 1-D slits and slit arrays[3] have also been proposed. Reassessment of the earlier data by new numerical studies and new measurements[4] have prompted a sharp downward revision of the enhanced transmission factor from ~ 1000 to ~ 10 and have motivated the development of a new model of surface wave excitation termed the composite diffracted evanescent wave (CDEW) model. This model builds a composite surface wave from the large distribution of diffracted evanescent modes (the inhomogeneous modes of the "angular spectrum representation" of wave fields[5] generated by a subwavelength feature such as a hole, slit, or groove when subjected to an external source of propagating wave excitation. The CDEW model predicts three specific surface wave properties. First, the surface wave is a composite or "wave packet" of modes each evanescent in the direction normal to the surface. The surface wave packet exhibits well-defined nodal positions spaced by a characteristic wavelength, second, the appearance of a phase delay of one-half pi with respect to the E-field of the external driving. source; and third, an amplitude decreasing inversely with distance from the launch site. We report on a series of experiments on very simple 1-D subwavelength surface structures designed to investigate these predictions and thus assess the validity of the model.

THE CDEW MODEL

The essential elements of the CDEW model can best be summarised with reference to Figure 1.

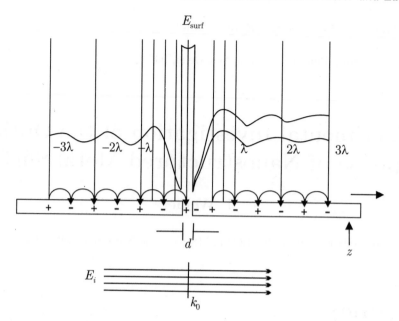

Figure 1. Essential elements of the CDEW model. The incoming plane wave E_i with $k_0 = 2\pi/\lambda_0$ in air ($n = 1$) is linearly polarised parallel to the plane of the structure and perpendicular to the slit of subwavelength width d. A fraction of the incoming light E_{surf} forms the composite diffracted wave in the $\pm x$ directions, and the blue trace (displaced above the surface for clarity) shows E_{surf}.

It is based on a solution to the 2-D Helmholtz equation in the near field and subject to the slab-like boundary conditions of a slit in an opaque screen. The form of the inhomogeneous or evanescent field on the $z = 0$ boundary is shown in Figure 1. At transverse displacements from the slit $|x| > d/2$, the evanescent component of the field at the surface $E_{\text{ev}}(x, z = 0)$ can be represented to good approximation by the expression,

$$E_{\text{ev}} \simeq \frac{E_i}{\pi} \frac{d}{x} \cos(k_{\text{surf}}x + \pi/2) \tag{1}$$

that describes a damped wave with amplitude decreasing as the inverse of the distance from the launching edge of the slit, a phase shift $\pi/2$ with respect to the propagating plane wave at the midpoint of the slit and a wave vector $k_{\text{surf}} = 2\pi/\lambda_{\text{surf}}$. The wavelength of the CDEW on the surface $\lambda_{\text{surf}} = \lambda_0/n_{\text{surf}}$ where n_{surf} is the surface index of refraction.

EXPERIMENT

When the composite evanescent wave encounters a surface discontinuity (a slit for example), a fraction of the surface wave is reconverted to a distribution of "homogeneous" or propagating modes $|k| = 2\pi/\lambda_0$ at the site of the slit. In a practical experiment, any real planar structure has two surfaces: an "input side" in the half-space $z < 0$, containing the incoming plane wave, and an "output side" in the half-space $z \geq 0$, containing the far-field propagating modes issuing from the output surface and a photodetector. Experiments can be carried out by fabricating subwavelength grooves on the input side, the output side or both. Figure 2 shows a schematic of the subwavelength structure we have used to investigate the amplitude and phase of the waves.

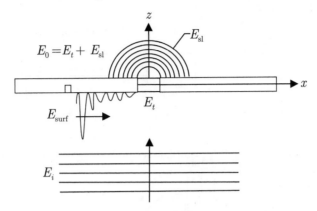

Figure 2. The incoming plane wave E_i impinges on the subwavelength slit (or hole) and a groove milled on the input side. The evanescent E_{surf} wave originates on the surface at a slit-groove distance x_{sg} and is indicated in blue. Incoming and surface wave interfere at the slit. The interference fringe is detected in the far field on the $z > 0$ side of the structure

Measurements of the optical response of the slit-groove and hole-groove structures were carried out using a home-built goniometer shown in Figure 3. We have carried out a series of measurements on simple 1-D structures to test the "signature" predictions of the CDEW model, viz. (1) a composite surface wave approximately represented by Eq. (1); (2) a phase shift of $\pi/2$ between the CDEW and the driving source plane wave, and (3) a wave amplitude that decreases inversely with distance from the launching groove.

The slit-groove structures were mounted facing the input side and exposed to plane-wave radiation from the focused TEM_{00} laser source. Measurements of light intensity on the output side in the far field, 200 mm from the plane of the structures, were carried out on the slit-groove structures using the goniometer setup shown in Figure 3. The results are shown in Figure 4. They show an oscillatory fringe pattern with amplitude damping out to a distance of 3 to 4 microns and maintaining an essentially constant amplitude from that point out to the distance limit of the measurements. As indicated in Figure 2, the fringe pattern results from interference between the mode directly propagating through the slit (hole) at the input side E_t and a surface wave originating from the single-groove structures E_{surf}. The wave E_{surf} is reconverted to a propagating mode at the slit or hole, and it is this propagating mode that interferes with E_t. The frequency and phase of the interference pattern is a function of the slit-groove optical path and any intrinsic phase shift of the surface wave itself. The normalised intensity I/I_0 of the superposition term is given by,

$$\frac{I}{I_0} = 1 + \eta_i^2 + 2\eta_i \cos\gamma_i \quad \text{with} \quad \eta_i = \frac{\alpha\beta}{\delta} \tag{2}$$

where $\alpha = E_{\text{surf}}/E_i$ is the fractional amplitude of the surface wave launched from the incoming field E_i at the groove site, and β is the further fraction of this surface wave reconverted to a propagating wave in the slit, $E_{sl} = \beta E_{\mathrm{surf}} = \beta E_i$. The fractional amplitude of the directly transmitted component E_t is δ and the phase difference γ_i between E_t and E_{sl} is the sum of two terms,

$$\gamma_i = k_{\mathrm{surf}} x_{sg(hg)} + \varphi_{\mathrm{int}}. \tag{3}$$

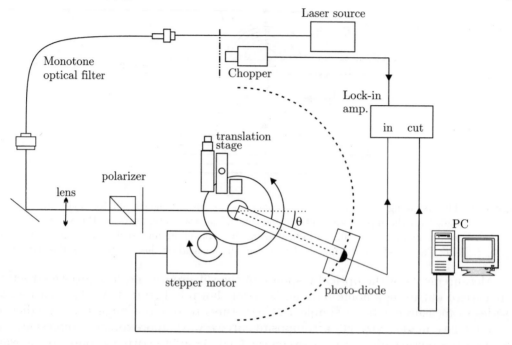

Figure 3. Goniometer setup for measuring far-field light intensity and angular distributions. A stabilised single mode CW diode laser, locked to a wavelength of 852 nm and modulated at 850 Hz by a chopper wheel, is injected into a single-mode fibre and focused onto the nanostructures mounted in a x-y translation stage as shown. A stepper motor drives the goniometer arm, and the chopped light intensity detected by the photodiode is fed to a lock-in amplifier. Output from the lock-in is registered by the PC that also drives the stepper motor. For the input-side experiments described here the detector was always positioned at $\theta = 0$

The first term $k_{\mathrm{surf}} x_{sg(hg)}$ is the phase accumulated by the surface wave propagating from the groove to the slit (or hole) and the second term φ_{int} is any phase shift intrinsic to the surface wave. The term φ_{int} includes the "signature" shift of the CDEW plus any phase shift associated with the groove width and depth.

RESULTS

Figure 4 presents a direct measure of the normalised amplitude damping with distance, $\eta_i = \eta_i(x)$ and the period and phase of the oscillations, from which the wavelength λ_{surf} of the surface wave, the phase φ_{int}, and the effective surface index of refraction n_{surf} can be determined. Analysis of the frequency spectrum of the fringe pattern for the slit structures results in the determination of a surface wavelength $\lambda_{\mathrm{surf}} = 819(811) \pm 8\,\mathrm{nm}$ and an effective surface index of refraction $n_{\mathrm{surf}} = 1.04(1.05) \pm 0.01$.

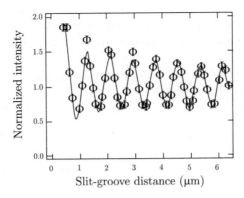

Figure 4. Noramalised far-field intensity as a function of slit-groove distance for a series of structures

DISCUSSION

How do these results compare to CDEW model? In the CDEW picture, the groove launches a surface wave on the input side of the silver film that is detected by interference with the directly transmitted wave through the hole or slit, in the far field, on the output side of the structure. The amplitude of this surface wave is predicted to damp as the inverse distance between the groove and the slit. Figures 4 shows an initial decrease in amplitude with increasing distance out to about 3-4 μm, but that the amplitude thereafter remains essentially constant. The constant amplitude beyond 3-4 μm is consistent with a persistent surface wave. Indeed we have recorded measurements (not presented here) of the surface wave persisting at least to $\sim$30 μm slit-groove distance. It is important to emphasize, however, that λ_{surf} and n_{surf} deviate significantly for those expected for a pure SPP on a plane silver surface. Interferometery measurements of the surface waves on "output side" slit-groove structures (not reported here) confirm the value of n_{surf}, and we believe that conventional, infinite-plane SPP theory is not adequate to explain these results.

The interpretation that emerges from these results is that the subwavelength groove originates persistent, long-range surface waves by a two-step process: (1) the incoming TM polarised plane wave scatters from the groove and generates in its immediate vicinity on the surface a broad, CDEW-like distribution of diffracted evanescent waves, and (2) this broad-band local surface "emitter" excites, within a distance of 3-4 μm, a long-range surface wave response. The near-term rapid amplitude decrease in the interference fringes of Figure 4 is evidence of this evanescent surface wave diffraction very near the groove. Persistent amplitude out to tens of microns is evidence for some kind of surface wave guided mode.

REFERENCES

1. T. W. Ebbesen, H. J. Lezec, H. F. Ghaemi, T. Thio and H. J. Wolff (1998), Extraordinary optical transmission through sub-wavelength hole arrays, Nature, 391, 667-669.
2. M. J. Treacy (2002), Dynamical diffraction explanation of the anomalous transmission of light through metallic gratings, Phys. Rev. B, 66, 195105-1–195105-11.
3. Q. Cao and P. Lalanne (2002), Negative role of surface plasmons in the transmission of metallic gratings with very narrow slits, Phys. Rev. Lett., 88, 057403-1–057403-4.

4. H. J. Lezec and T. Thio (2004), Diffracted evansecent wave model for enhanced and suppressed optical transmission through subwavelength hole arrays, Opt. Express, 12, 3629–3651.

5. L. Mandel and E. Wolf (1995), Optical Coherence and Quantum Optics, 109-120, Cambridge University Press, Cambridge, England.

Atoms and Molecules in Laser and External Fields
Editor: Man Mohan

Attochemistry is Coming On

Andre D. Bandrauk[1], Szczepan Chelkowski[1] and Gennady L. Yudin[1,2]

[1] Laboratoire de Chimie Théorique, Université de Sherbrooke, Sherbrooke,
Québec J1K 2R1, Canada
[2] National Research Council of Canada, Ottawa, Ontario K1A 0R6, Canada

I. INTRODUCTION

Advances in current laser technology allow experimentalists to have access to new laser sources for examining molecular structure and dynamics on the natural time scales for atomic (vibrational) motion in molecules, the femtosecond (fs, 10^{-15} s) [1]. These new sources are ultrashort ($t_p < 10$ fs) and intense ($I \geq 10^{14}$ W/cm^2) [2]. Such pulses lead to a new regime of laser-matter interaction, the nonlinear, non-perturbative regime. Thus since the atomic unit (a. u.) of electric field strength $E_0 = e/a_0^2 = 5 \times 10^9$ V/cm ($a^0 = 1$ a. u. $= 0.0529$ nm) corresponds to an a. u. of intensity, $I_0 = cE_0^2/8\pi = 3.5 \times 10^{18}$, laser intensities approaching I_0 have led to the discovery of new nonlinear phenomena in laseratom experimental and theoretical studies [2,3], such as *above-threshold ionization* (ATI), the multiphoton equivalent of single-photon ionization, and *high-order harmonic generation* (HHG).

One of the seminal ideas emanating from theories of intense laser a atom interactions is a simple quasistatic model of electron tunnel ionization [2] followed by recollision of the ionized electron with its parent ion [4]. The returning energy of the electron at the nuclei can be shown to satisfy a simple recombination law, which explains the maximum energy of harmonics emitted by such a recolliding elecron with its parent ion,

$$N_m \hbar \omega = I_p + 3.17 U_p. \tag{1}$$

N_m is the maximum harmonic order, I_p is the ionization potential, and U_p is the ponderomotive or oscillatory energy of an electron of mass m in a pulse of frequency ω and intensity I,

$$U_m[\text{a.u.}] = \frac{I}{4m\omega^2} = 3.4 \times 10^{21} I[\text{W/cm}^2]\lambda^2[nm]. \tag{2}$$

The corresponding field induces displacement of such an ionized electron is given by the expression

$$\alpha[\text{a.u.}] = \frac{eE}{m\omega^2} = 2.4 \times 10^{-12}(I[\text{W/cm}^2])^{1/2}\lambda^2[nm]. \tag{3}$$

Thus at an intensity $I = 1014$ W/cm^2 and wavelength $\lambda = 1064$ nm, $U_p = 10.5$ eV and $\alpha = 1.2$ nm. In molecules electron collisions can occur with neighboring ions and this generally leads to even lager harmonic orders with photon energies of $6U_p$ and even up to $126U_p$, thus extending the atomic recollision maximum energy cut-off law 1 [5].

Using ultrashort intense laser pulses ($t \leq 5fs$) results in continuum HHG in the regime of the cut-off law 1 from which by selecting a slice of the spectrum one can synthesize "attosecond" (asec) pulses. First experimental results on the production and measurements of sub-fs, asec, pulses from high-intensity ionization of Ar atoms was reported in the last few years [6, 7]. This opens up new possibilities for studying and controlling [8] electron dynamics in atoms and molecules on the electron, i. e., asec time scales. We have recently shown that asec pulses can be used to measure the absolute electric field of intense few-cycle laser pulses [9]. Furthermore this leads to "asec control" of ionization and thus to increase HHG intensities allowing to create new single X-ray or VUV pulses as short as 250 asec [10]. Even shorter asec pulses have been predicted from exact numerical solution of coupled *Maxwell-time-dependent Schrödinger equations* (TDSE) for the molecular ion H_2^+ which represents a first complete treatment of photon-electron-nuclear dynamics [11]. We were able to demonstrate from solutions of such TDSE simulations how pump-probe experiments of the photoionization of H_2^+ using asec pulses can allow one to study the evolution of the movement of localized electron wave packets, thus leading to new asec molecular spectroscopy [12]. Such experiments should now be feasible with the recent discovery of very large HHG from ions themselves at very high intensities [13, 14], thus extending asec pulse into the soft X-ray region.

Recent advances in ultrashort laser-pulse technology enables experimentalists to synthesize and characterize ultrashort pulses in the attosecond time regime [15-20]. The development of femtosecond technology has been an interplay between the needs of science, demanding ever faster measurements, and technology producing ever shorter pulses [21]. Chirped pulses play an important role in technology, transform-limited pulses are used for measurement. Attosecond technology is a radical departure from the ultrafast science that proceeded it, see review [22]. Chirp is inherent in the attosecond generation process [4] except near the cut-off frequency [16, 17]. Pulses with the broadest phased bandwidth will have substantial chirp [18].

Measuring electron dynamics in atoms and molecules [23-25] is a natural initial target for attosecond science. When measurement is via photoelectrons, chirped pulses allow the same resolution limit as transform-limited pulses of the same bandwidth. What is essential is that the measurements are differential. That is, the photoelectron spectrum is resolved in momentum and in space.

Our method is based on broad bandwidth photoionization. Single-photon ionization from a stationary state produces photoelectrons that contain information on the optical pulse and the phase and structure (through the transition moment) of the initial state. If the state is known, measurement of the photoelectron spectra produced in the presence of a phased infrared field fully characterizes the attosecond pulse (see references in review [19] and more recent papers [26-28]); if, on the other hand, the attosecond pulse is known, measurement of photoelectron spectra produced from two or more coherently excited states, determines all information on the associated bound-state wave packet.

Photoionization is related to a proposal to measure bound-state electron dynamics during the attosecond generation process itself [23]. Attosecond single-photon ionization has three advantages over strong-field techniques [24, 25]. Specifically, we can: (i) measure a bound electron state that is not perturbed by a strong laser field, (ii) correlate photoelectrons to ion fragments so that we can specify the molecular internal states and alignment direction, and (iii) precisely take into account the Coulomb potential in the molecular continuum.

II. ATTOSECOND MOLECULAR PHOTOELECTRON SPECTROSCOPY

A.   General Results

We assume that a pump pulse has prepared a coherent superposition of two bound states,

$$\Psi(\boldsymbol{r}, t) = \alpha_1 \Psi_1(\boldsymbol{r}, t) + \alpha_2 \exp(-i\beta)\Psi_2(\boldsymbol{r}, t) \tag{4}$$

where $\alpha_{1,2}^2$ are their populations ($\alpha_1^2 + \alpha_2^2 = 1$) and β is an adjustable initial phase that depends on the excitation scheme. The coherently prepared state is the simplest wave packet that can be formed. The pump pulse does not need to be short.

Now we consider single photon photoionization induced by a probe attosecond x-ray pulse. The probe pulse is precisely timed with respect to the phase of the pump pulse. Phasing occurs naturally in attosecond technology if the initial wave packet is excited by the fundamental pulse or its harmonics. We define the vector potential of such a pulse as a linearly chirped Gaussian

$$\boldsymbol{A}(t) = e(A_0/2)\exp[-i\Omega t - (t - t_0)^2/2(1 - i\xi)\tau^2], \tag{5}$$

where Ω is the central frequency of the attosecond pulse, e, A_0, and t_0 are the linear polarization vector, amplitude, and peak of the pulse. Positive dimensionless chirp x corresponds to the instantaneous frequency increasing with time. The pulse duration is Eq. (5) describes a pulse whose bandwidth is invariant to the pulse chirp.

The photoionization amplitude for the coherently coupled state (4) and three-fold differential angle resolved photoelectron spectrum are given by

$$M_{coh} = M_o \left[\Upsilon_1 M^{(1)} + \Upsilon 2 e^{-i[\delta(t_o) + \phi(\xi, p)]} M^{(2)} \right], \tag{6}$$

$$\frac{d^3 P_{coh}(p, \theta, \varphi)}{dp d\Omega_e} \equiv S(p, \theta, \varphi) = p^2 |M_{coh}|^2, \tag{7}$$

where $M_0 = A_{0\varepsilon\tau}[\pi(1 - i\xi)/2]^{1/2}$, $M^{(1)}$ and $M^{(2)}$ are the photoionization amplitudes for two bound states in the coherent superposition (4), ε is a complex phase factor, $|\varepsilon| = 1, \gamma_{1,2} = \alpha_{1,2}\exp(-\omega_{1,2}^2\tau^2/2), \omega_{1,2} = p^2/2 + I_p^{(1,2)} - \Omega, I_p^{(1,2)}$ are the ionization potentials of states $\Psi_1(r; t)$ and $\Psi_2(r; t)$, and $d\Omega_e = \sin\theta d\theta d\varphi$ is the solid angle (unless stated otherwise, we use atomic units $e = m_e = \hbar = 1$).

Eq. (6) and Figure 1 show that there are two contributions to the photoelectron spectrum. The bandwidths $\Delta_{1,2}$ of electron wave packets in Figure 1 are defined by functions $\gamma_{1,2}$ in the amplitude (6). It is of importance that both $\gamma_{1,2}$ and $\Delta_{1,2}$ do not depend on linear chirp of the x-ray pulse (for example, the transform-limited $\tau_{FWHM} = 24.19$ as (i.e., 1 a. u.) pulse and chiped $\tau_{FWHM} = 250$ as pulse at $\xi \approx 10$

generate electron wave packets with the same bandwidths). For a broad-bandwidth pulses a photoelectron wave packets overlap and this spectral overlapping is the same for a given transform-limited pulse ($\xi = 0$) and its chirped modifications.

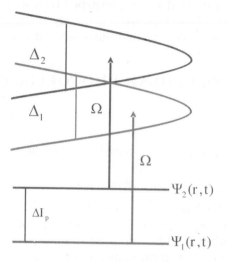

Figure 1. A sketch of an electron wave packet generated by photoionization of a coherent superposition of electronic states. The bandwidths $\Delta_{1,2}$ and spectral overlapping of wave packets do not depend on linear chirp of x-ray pulse

The resolution limit of a wave packet measurement is independent of the chirp. The interference, however, is different in details and this difference offers a new possibility for using chirped pulses to make single-time delay attosecond dynamics measurements. The interference structure is determined by the full phase shift between amplitudes in (6). This phase is universal and does not depend on structure of the photoionization amplitudes $M^{(1,2)}$. However it depends on the chirp. The time-dependent phase shift $\delta(t) = \beta + \Delta I_p t$ parameterizes the bound electron motion. The chirp-dependent part of the full phase shift [25]

$$\phi(\xi, p) = \xi \tau^2 \Delta I_p [p^2/2 + I_p^{(av)} - \Omega] \tag{8}$$

describes the chirped interference inside the photoelectron spectrum. $\Delta I_p = I_p^{(1)} - I_p^{(2)}$ is the level separation which depends on R in molecular case and $I_p^{(av)} = (I_p^{(1)} + I_p^{(2)})/2$ is the average ionization potential of coherently coupled electronic states.

Both phase factors are periodic. The factor $\exp[-i\delta(t_0)]$ which corresponds to electron "hopping" [12, 24] is the periodic function of time with the period $T_0 = 2\pi/\Delta I_p$ and the factor $\exp[-i\phi(\xi, p)]$ is periodic function of the photoelectron energy $E_p = p^2/2$ with the period [25]

$$E_0 = T_0 = |\xi|\tau^2 = 2\pi/|\xi|\tau^2 \Delta I_p. \tag{9}$$

Experimentally it may be more useful to measure the momentum asymmetry than the momentum spectrum because the momentum asymmetry is self calibrating. In analogy with the integral asymmetry [32] (see also [24]), the absolute (Δ) and normalized (Δ_N)

differential asymmetries are defined as

$$\Delta(p,\theta,\varphi) = S(p,\theta,\varphi) - S(p,\pi-\theta,\varphi), \tag{10}$$

$$\Delta_N(p,\theta,\varphi) = \frac{S(p,\theta,\varphi) - S(p,\pi-\theta,\varphi)}{S(p,\theta,\varphi) + S(p,\pi-\theta,\varphi)}. \tag{11}$$

The asymmetry (10) is independent of the azimuthal angle φ since the molecule is aligned along the z-axis.

B. Attosecond Photoelectron Spectra

So far, our description is completely general. However, for illustrative purposes it is useful to select a specific case. As a theoretically tractable example that capture the essence of our approach, we consider the H_2^+ molecular ion with or without a coherent superposition of two electronic states, $\sum_g 1s$ (state $|1\rangle$) and $\sum_g 2p_0$. The electron and nuclear time scales in molecule remain well separated so the nuclei are considered fixed. We assume that the molecular ion is aligned. Alignment can be effectively achieved experimentally for unstable molecules since, following ionization, the molecular fragments determine the alignment with high accuracy [29-31]. In our three-dimensional model, the molecular ion and electric field of the attosecond pulse are aligned along the z axis in Cartesian space.

We describe the initial H_2^+ wave functions as a linear combinations of atomic orbitals centered on each nucleus:

$$\Psi_b(r) = \sum_i c_i \phi_i^{(1)}(r_1) + \sum_j c_j \phi_j^{(2)}(r_2), \tag{12}$$

where vectors $r_{1,2} = r \pm R/2$, r is the electron coordinate, and R is directed from the nucleus 1 to the nucleus 2. The number of the atomic basis functions in the linear combination and their variational Slater parameters $a = 1 = Z_{eff}$ depend on internuclear distance R. In homonuclear molecules at large R Eq. (12) can be reduced to linear combination of two identical atomic orbitals:

$$\Psi_b^{\pm}(r) \approx c_{\pm}[\phi(r_1) \pm \phi(r_2)]. \tag{13}$$

We calculate the molecular photoionization amplitude using the final two-center Coulomb wave function. For the simplest choice (13), the molecular ionization transition amplitudes are determined by atomic amplitudes:

$$M_{mol}^{\pm} \approx \chi_{\pm} c_{\pm} M_{at}, \tag{14}$$

where $\chi_{\pm}$ are the molecular interference factors [24, 25]:

$$\chi_{\pm} = N_p[\exp(ipR/2)G(-R) \pm \exp(-ipR/2)G(R)]. \tag{15}$$

$N_p = \exp(\pi/2p)\Gamma(1+i/p)$ is the normalization factor and $G(R)$ is the confluent hypergeometric function, $G(R) = F(i/p; 1; i(pR + pR))$.

For isolated initial molecular states, the attosecond photoionization spectra are displayed in Figures 2 and 3. The appropriate dipole atomic photoionization amplitudes are calculated in [24, 25]. The spectra are generated by the chirped $\tau_{FWHM} = 100\sqrt{1+\xi^2}$ as pulse. The angle between the vectors e and p is $\theta_e = 30°$, $\Omega = 100\,\text{eV}$. With changing of R the molecular ionization potentials slightly change, but interference patterns is defined mostly by the values of pR and pR in the interferences factors (15).

The photoelectron signal for ionization of the $\sum_g 1s$ initial state at $R = 6.8$ (Figure 2) and for the $\sum_g 2p_0$ initial state at $R = 8.0$ (Figure 3) are suppressed due to destructive molecular interference. The insets correspond to the different interference factors, χ_+ in the case of $\sum_g 1s$ initial state and χ_- in the case of $\sum_g 2p_0$ initial state. Note that the spectra are very sensitive to all parameters of the attosecond pulse and molecular geometry.

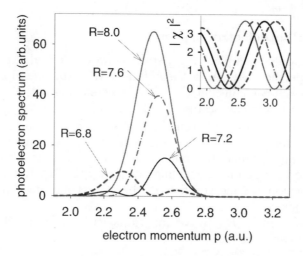

Figure 2. Molecular attosecond photoionization spectra generated by the chirped $\tau_{FWHM} = 100\sqrt{1 + \xi^2}$ as pulse at different large internuclear distances. The angle between the vectors e and p is $\theta_e = 30°$, $\Omega = 100$ eV. The initial molecular state is $\sigma_g 1s$ and the interference factor $\chi = \chi_+$

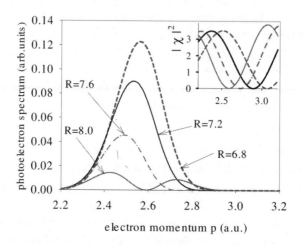

Figure 3. Molecular attosecond photoionization spectra generated by the chirped $\tau_{FWHM} = 100\sqrt{1 + \xi^2}$ as pulse at different large internuclear distances. The angle between the vectors e and p is $\theta_e = 30°$, $\Omega = 100$ eV. The initial molecular state is $\sigma_g 2p_0$ and the interference factor $\chi = \chi_1$

The absolute molecular asymmetries for the case of $\sum_g 1s + \sum_u 2p_0$ coherently coupled states are displayed in Figure 4. The populations of coherently coupled states are equal, $(\alpha_1^2 = \alpha_2^2)$ central photon energy $\Omega = 100$ eV, and angle between polarization and photoelectron momentum $\theta = \pi/3$. The photoelectron signal and, therefore, the absolute asymmetry at $R = 12$ is suppressed due to destructive molecular interference. The minimum value of $|\chi_+|^2$ in Figure 4 is in the range of the photoelectron momenta.

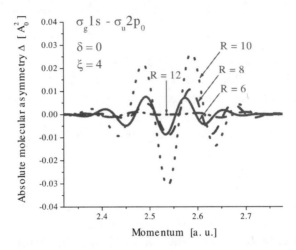

Figure 4. Absolute molecular asymmetries for the case of $\sigma_g 1s + \sigma_u 2p_0$ coherently coupled states. The populations of coupled states are equal, $(\alpha_1^2 = \alpha_2^2)$, central photon energy $\Omega = 100$ eV, chirp $\xi = 4$, and angle between polarization and photoelectron momentum $\theta = \pi/3$

III. CONCLUSION

In conclusion, attosecond pulses, because of their large energy bandwidth and high frequency, project bound-state electron wave packets into the continuum. There, the wave packet can be fully characterized through its angle-resolved photoelectron spectrum. In practice, current attosecond laser sources are not intense enough and operate at a low repetition rate, making a full reconstruction impractical. However, they are perfectly adequate to resolve attosecond dynamics by measuring the changing momentum asymmetry of the spectrum [24, 25]. This would allow attosecond electron dynamics experiments that are analogous to vibrational dynamics experiments, which have been so influential in photochemistry [1]. As much higher repetition rate systems are developed [33] full characterization of both the spatial and temporal structures of electron wave packets will become possible.

Long chirped attosecond pulses can measure attosecond time scale electron dynamics. The dynamics can be measured just as effectively as if the pulses were transform-limited. Pump-probe time delay spectroscopy is possible, but not necessary if chirped pulses are used [25]. The oscillations in the photoelectron spectra are described by the universal phase shift between the interfering photoionization amplitudes. The phase shift allows us to read the dynamics at a single time delay. The differential normalized asymmetry of the photoelectrons reaches 100% [25]. All electrons in some momentum windows are sent in a given direction. The direction changes with time delay.

Until now figure-of-merit of attosecond pulses has been the pulse duration. Our results imply that the emphasis of attosecond technology should shift from producing transformlimited pulses to producing broad bandwidth pulses. The pulses can be chirped or unchirped. All that is needed is that the chirp is well characterized.

Acknowledgments

We are grateful to P. B. Corkum, M. Yu. Ivanov and F. Krausz for valuable discussions.

REFERENCES

1. A. H. Zewail, J. Phys. Chem. A, 104 (2000), 5660.
2. T. Brabec and F. Krausz, Rev. Mod. Phys., 72 (2000), 545.
3. M. Gavrila, Atoms in Intense Laser Field, Academic Ptress, New York, 1992.
4. P. B. Corkum, Phys. Rev. Lett., 71, (1993) 1994.
5. A. D. Bandrauk and H. Yu, J. Phys. B, 31 (1998), 4243; Phys. Rev. A, 59 (1999), 539.
6. M. Nentschel et al., Nature (London), 414 (2001), 509.
7. R. Kienberger et al., Science, 297 (2002), 1144.
8. A. D. Bandrauk, Y. Fujimura and R. Gordon (editors), Laser Control and Manipulation of Molecules, ACS Symposium Series, Vol. 821, Washington, DC, 2002.
9. A. D. Bandrauk, S. Chelkowski and H. S. Nguyen, Phys. Rev. Lett., 89 (2002), 283903.
10. A. D. Bandrauk and H. S. Nguyen, Phys. Rev. A, 66 (2002), 031401.
11. A. D. Bandrauk, S. Chelkowski and H. S. Nguyen, J. Mol. Struct., 735C (2004), 203.
12. A. D. Bandrauk, S. Chelkowski and H. S. Nguyen, Int. J. Quant. Chem., 100 (2004), 834.
13. E. A. Gibson et al., Phys. Rev. Lett., 92 (2004), 033001.
14. S. Seres et al., Phys. Rev. Lett., 92 (2004), 163002.
15. M. Drescher et al., Nature (London), 419 (2002), 803.
16. A. Baltuka et al., Nature (London), 421 (2003), 611.
17. R. Kienberger et al., Nature (London), 427 (2004), 817.
18. Y. Mairesse et al., Science, 302 (2003), 1540.
19. P. Agostini and L. F. DiMauro, Rep. Progr. Phys., 67 (2004), 813.
20. R. Lpez-Martens et al., Phys. Rev. Lett., 94 (2005), 033001.
21. J.-C. Diels and W. Rudolph, Ultrashort Laser Pulse Phenomenon: Fundamentals, Techniques and Applications on a Femtosecond Time Scale, Academic Press, Boston, 1996.
22. J. Levesque and P. B. Corkum, Can. J. Phys., to be published.
23. H. Niikura, D. M. Villeneuve and P. B. Corkum, Phys. Rev. Lett., 94 (2005), 083003; H. Niikura et al., J. Mod. Optics 52 (2005), 453.
24. G. L. Yudin et al., Phys. Rev. A, 72 (2005), 051401(R).
25. G. L. Yudin, A. D. Bandrauk and P. B. Corkum, submitted.
26. J. Itatani et al., Laser Phys., 14 (2004), 344.
27. Y. Mairesse and F. Quere, Phys. Rev. A, 71 (2005), 011401(R); F. Quere, Y. Mairesse, and J. Itatani, J. Mod. Optics, 52 (2005), 339.
28. E. Cormier et al., Phys. Rev. Lett., 94 (2005), 033905.
29. H. Niikura et al., Nature (London), 417 (2002), 917; 421 (2003), 826.

30. J. Ullrich et al., Rep. Prog. Phys., 66 (2003), 1463 ; C. Dimopoulou et al., J. Phys. B, 38 (2005), 593.

31. T. Weber et al., Phys. Rev. Lett. 92 (2004), 163001; Nature (London), 431 (2004), 437; A. L. Landers et al., Phys. Rev. A, 70 (2004), 042702.

32. A. D. Bandrauk, S. Chelkowski and N. H. Shon, Phys. Rev. Lett., 89 (2002), 283903; Phys. Rev. A, 68 (2003), 041802(R).

33. R. J. Jones et al., Phys. Rev. Lett., 94 (2005), 193201.

Atoms and Molecules in Laser and External Fields
Editor: Man Mohan
Copyright © 2008, Narosa Publishing House, New Delhi, India

Enhanced Optical Response in Multi-Atom Ensembles

M. Macovei, J. Evers and C. H. Keitel

Max-Planck Institute for Nuclear Physics, Saupfercheckweg 1, D-69117 Heidelberg, Germany

I. INTRODUCTION

Multi-particle interactions [1-11] substantially enhance the optical properties of large atomic samples. Interest in collective phenomena has grown recently because of their potential advantages in mesoscopic systems [12], quantum information processing [13], Bose-Einstein condensates [14] or in rapid control of collective quantum dynamics [15].

In this paper, we present some of our recent investigations on the manipulation of laser-driven multiatom few-level ensembles. In a mixture of two different two-level ensembles we demonstrate the possibility of creating atomic media with large index of refraction without absorption. We find quantum features in the *electromagnetic field* (EMF) emitted by a collection of three-level ladder radiators that interacts with a classical bath only. As applications, these vacuum-correlated multi-particle systems can be used for building switching devices, quantum filed generators or amplifiers.

The paper is organized as follows. In Section II we analyze the possibilities of creation of a high refractive index medium in a mixture of two two-level ensembles. The next Section III describes the non-classical properties of the EMF radiated by a sample of three-level emitters that interacts with incoherent fields. Finally, the results are summarized in Section IV.

II. ENHANCING THE REFRACTIVE INDEX VIA LASER-MEDIATED COLLECTIVITY

We show here that the mutual interactions of nearby atoms via quantum fluctuations of the surrounding electromagnetic field [1-5] are suitable to generate transparent media with large indices of refraction of order of magnitude 10, or high dispersion of arbitrary sign.

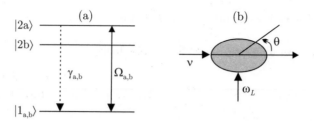

Figure 1. Schematic diagram depicting means of controlling refractive properties of two laser-driven atomic two-level ensembles. (a) The involved energy levels of the two two-level atomic systems, spontaneous decay rates, and Rabi frequencies are denoted with **a** and **b**. (b) The index of refraction and thus the deviation angle θ of a weak probe field (v) may be substantially and rapidly manipulated by an applied laser field (ω_1)

In particular we point out that the effects may be considerably larger than for the same number of independent atoms and may be set up on a time scale which is inversely proportional to the number of atoms in the sample [15, 16]. Furthermore, the group velocity of a weak probe field propagating through the strongly driven two-level medium may be substantially slowed down or accelerated.

For this purpose, we consider an atomic system consisting of two ensembles of two-level atoms, numbered N_a and N_b, with densities of order $10^{12} - 10^{14}$ cm^{-3}, and with somewhat different transition frequencies $\{\omega_a, \omega_b\}$, and interacting with a single moderately strong laser field (see Figure 1). The corresponding Rabi frequencies are $\{2\Omega_a, 2\Omega_b\}$ and spontaneous decay of all closely spaced atoms occurs via interaction with a common electromagnetic field reservoir with rates $\{2\gamma_a,\ 2\gamma_b\}$ from excited states $\{|2_a\rangle, |2_b\rangle\}$, respectively. In order to treat the atoms uniformly we suppose that $\{L/c, \Omega_{a,b}^{-1}\}$ where L, c and τ are the largest dimension of the sample, the light velocity and the collective decay time, respectively. For a pencil shaped sample with length $L \sim 5\lambda$, transverse area $S \sim 2\lambda^2$, $\sim 10^{-4}$ cm, $\gamma \sim 10^7$ Hz, and $N \sim 10^3$ we estimate a collective decay time, $\tau_s \sim 2L/(\lambda\lambda N)$, of about 10^{-9}s. When scanning the composed atomic sample with, say for instance, a dye or diode laser, and depending on the resonance condition for each kind of atom andthus the employed laser frequency, generally one or the other atomic species may dominate the finalsteady-state collective behavior. In what follows, we neglect the cross-correlations between the atoms **a** and **b**, and thus restrict ourselves to the case $N_{\gamma_{a,b}} \bullet |\omega_a - \omega_b| \bullet \widetilde{\Omega}_{a,b}$. Therefore, collective interactions via the environmental vacuum occur between atoms possessing identical transition frequencies, i.e. not different ones coupling with distinct parts of the vacuum.

We proceed by calculating the refractive properties of a very weak field probing the strongly driven atomic sample. The linear susceptibility $\chi(v)$ of the probe field, at frequency v, can be represented in terms of the Fourier transform of the average value of the two-time commutator of the atomic operator as:

$$\chi(v) = i \sum_{j \in \{a,b\}} (d_j^2/\hbar V_j) \int_0^\infty d\tau \exp(iv\tau)\langle [S_-^{(j)}(\tau), S_+^{(j)}]\rangle_s$$

[17,18]. The collective atomic operators $S_-^{(j)} = |1_j\rangle\langle 2_j|$, $S_+^{(j)} = |2j\rangle\langle 1_j|$ satisfy the standard commutation relations for the su(2) algebra. In the intense-field limit $\Omega_1 \gg {}_iN_i$, the susceptibility $\chi(v)$ transforms into the dressed-state picture, $|\Psi_j^{(i)}\rangle$ for $\{i = a, b\}$, and $\cot 2\theta_i = \Delta_i/2\Omega_i$, $\{i \in a, b\}$ via.

$$|1_i\rangle = |\Psi_1^{(i)}\rangle \cos\theta_i + |\Psi_2^{(i)}\rangle \sin\theta_i, |2_i\rangle = -|\Psi_1^{(i)}\rangle \sin\theta_j + |\Psi_2^{(i)}\rangle \cos\theta_i.$$

In the secular approximation [15,16], Figure 2 depicts the steady-state dependence of the linear susceptibility with respect to the strong laser detunings while keeping fixed probe-field frequencies. Strong gain as well as strong positive or negative dispersion with zero absorption are then feasible. The interpretation of these results is straightforward via a dressed-state analysis. When $\nu - \omega_a = -2\Omega$ (see Figure 2(a)) the probe field is at exact resonance with the dressed-state transition $|\Psi_1^{(a)}\rangle \leftrightarrow |\Psi_2^{(a)}\rangle$. If $\Delta_n/(2\Omega) < 0$, the dressed - state population is placed in the dressed - states $|\Psi_2^{(i)}\rangle$ and thus, the probe field is absorbed.

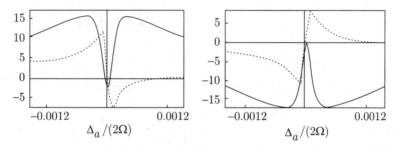

Figure 2. The steady-state dependence of the linear susceptibility χ (in units of $\bar{N}d^2/\gamma\hbar$ as a function of $\Delta_a/(2\Omega)$). The solid and dashed lines correspond to the real and imaginary parts of, respectively. Here $\nu - \omega_a = -2\Omega(a)$, and $\nu - \omega_a = -2\Omega(b)$ while $N = 1500, 2\Omega/(N_\gamma) = 10, \Delta\omega(N_\gamma) = 0.1$, and $r/\gamma = 0.3$

And thus, the probe field is absorbed. Increasing further Δ_a, i.e $\Delta_a/(2\omega) > 0$, the dressed-state population a a transfers to and the probe field is amplified. The second ensemble contributes here to a strong shift of the susceptibility resulting in zero absorption with large dispersive features. In particular, close to vanishing absorption, the index of refraction $n(\nu) \approx \sqrt{1 + \chi(\nu)}$ for atomic densities of order 10^{14} cm^{-3} takes values larger than 8 [see Figure 2(a) near $\Delta_a/(2\Omega) \sim 10^{-3}$]. However, without collective effects, i.e. a non- a interacting ensemble of $N_a = N_b = 1500$ atoms, the index of refraction would only be close to unity at the point of vanishing absorption. The index of refraction n can be further enhanced by increasing the atomic densities, but then the atoms would be so close to each other that short-range dipole-dipole interactions need to be taken into account. Also the refractive index may take values below unity in our system. Note that the collisional damping (with a rate r) does not affect considerably the collective steady-state behavior, and its influence can be balanced by increasing the number of atoms.

We demonstrate further that collections of two-level atoms are suitable for switching between strongly accelerating or slowing down of a weak probe pulse traversing through the driven two-level media. The light group velocity can be estimated from the following expressions: $1/v_g = n_g/c = dk(v)/dv$, where $k = n(v)v/c$. For $\bar{N}d^2/(\hbar\mu) \sim 0.1/ \sim 10^8, n_g$ may reach values of order of 107 of either sign. Thus, by g properly choosing the external parameters one can arrive at rather low subluminal or large superluminal group velocities.

III. QUANTUM FEATURES OF LIGHT EMITTED BY A COLLECTION OF ATOMS IN A THERMAL ENVIRONMENT

This section describes the interaction of an ensemble of N identical non-overlapping three-level ladder atoms with a thermal environment. The emitters are located within a volume with linear dimensions smaller than the relevant emission wave-lengths. The excited atomic levels, during the process $|1\rangle \rightarrow |2\rangle \rightarrow |3\rangle$ spontaneously decay to the state $|2(|3)$ with a decay rate $2_{\gamma_1}(2_{\gamma_2})$. The only external driving is via the 1 2 surrounding bath, which induces transitions among the atomic levels. In the usual mean-field, Born- Markov, and rotating-wave approximations, the system is described by the following master equation [19]:

$$\dot{\rho}(t) = -\gamma_1(1+\bar{n}_1)[S_{12}, S_{21}\rho] - \gamma_2(1+\bar{n}_2)[S_{23}, S_{32}\rho]$$
$$- \gamma_1\bar{n}_1[S_{21}, S_{12}\rho] - \gamma_2\bar{n}_2[S_{32}, S_{23}\rho] + H.c.$$

Here, an overdot denotes differentiation with respect to time. $\bar{n}_i = [\exp(\beta\hbar\omega_{i,j+1})-1]^{-1}$ is the mean thermal photon number at transition frequency $\omega_{i,i+1} = \omega_i - \omega_{i+1}$ and for temperature T, where with $\beta = (k_B T)^{-1}$ with k_B as the Boltzmann constant. The diagonal elements $P_{nm} = \langle N, n, m|\rho_{ss}|N, n, m\rangle$ of the steady-state density operator ρ_{ss} evaluate to [19].

$$P_{nm} = (1 - \eta_2)\eta_1^n\eta_2^m \left[\frac{1 - (\eta_1\eta_2)^{N+1}}{1 - \eta_1\eta_2} - \eta_2^{n+1}\frac{1 - \eta_1^{N+1}}{1 - \eta_1}\right]^{-1},$$

with $\eta_i = \bar{n}_i/(1+\bar{n}_i), (i \in \{1, 2\})$. Any atomic expectation value can easily be evaluated by using this expression. We now turn to our main interest in this study, the coherence properties of the collective fluorescence light generated on the two transitions $|1\rangle \rightarrow |2\rangle$ and $|2\rangle \rightarrow |3\rangle$. The second order coherence function is defined as

$$g_{ij}^{(2)}(\tau) = \frac{\langle J_i^+(t)J_j^+(t+\tau)J_j(t+\tau)J_i(t)\rangle}{\langle J_i^+(t)J_i(t)\rangle\langle J_j^+(t)J_j(t)\rangle},$$

where $i, j \in \{1, 2\}$ with $J_1 = S_{21}$ and $J_2 = S_{32}$. The quantity can be interpreted as a measure for the probability for detecting one photon emitted on transition i and another photon emitted on transition j with time delay τ. $g_{ij}^{(2)}(0) < 1$ characterizes sub-Poissonian, $g_{ij}^{(2)}(0) > 1$ super-Poissonian, and $g_{ij}^{(2)}(0) = 1$ Poissonian photon statistics. $g_{ij}^{(2)}(\tau) > g_{ij}^{(2)}(0)$ is the condition for photon anti-bunching, whereas $g_{ij}^{(2)}(\tau) < g_{ij}^{(2)}(0)$

is the condition for photon anti-bunching, whereas $g_{ij}^{(2)} < g_{ij}^{(2)}(0)$ means bunching. We further define super-bunching as bunching with $g_{ij}^{(2)}(0) > 2$. More specific, correlation functions with $i = j$ describe the photon statistics of the fluorescence light emitted on a single atomic transition, and $g_{i\neq j}^{(2)}(0)$ the cross correlations between the photon emission on two different transitions.

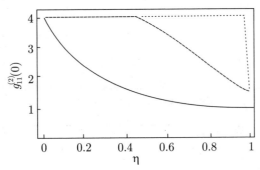

Figure 3. The second-order correlation function $g_{11}^{(2)}(0)$ versus the parameter $\eta \equiv \eta_1 = \eta_2$. Solid, long-dashed, and short-dashed curves are for $N = 2, 10$ and 300, respectively

In the following, we discuss the light properties emitted during $|1\rangle \rightarrow |2\rangle$ transitions. For a weak bath the emitted fluorescence light is completely incoherent since $g_{11}^{(2)}(0) > 2$ showing super-Poissonian statistics of photons because $\lim_{n\to 0} g_{11}^{(2)}(0) = 4$, for any N. Increasing the strength of the thermal field for $N > 2$, the radiating field changes from incoherent to partially coherent properties because $1 < g_{11}^{(2)}(0) < 2$ Interesting behavior occurs for $N = 2$. In this case, the emitted photons show super-Poissonian statistics for a weak bath and sub-Poissonian, i.e. quantum features, for a strong thermal bath (see Figure 3). The statistics of the emitted photons in the collective limit remains unchanged up to very intense temperatures when it switches from completely incoherent to almost coherent

$$\lim_{n\to 1} g_{11}^{(2)}(0) = 8(N-1)(N+4)/[5N(N+3)] \rightarrow 8/5, \text{ as } N \rightarrow \infty$$

The fluorescence field emitted by a few atom-system $(N \bullet 3\bullet)$ on the $|2\rangle \rightarrow |3\rangle$ atomic transition is partially coherent because $g_{22}^{(2)}(0) < 2$ (see Figure 4). For $N = 2$ the coherence function $g_{22}^{(2)}(0)$ changes from unity (coherent light) to values less than one demonstrating, thus, quantum features of the emitted radiation. The light statistics of a large sample behaves as follows: for a weak bath $(\eta_1 = \eta_2 < 1)$ it is incoherent as $\lim_{N\to\infty} g_{22}^{(2)}(0) = 2$, indicating photon bunching, while for an intense thermal reservoir (i.e. $\eta_1 = \eta_2 = 1$) it is partially coherent since $\lim_{n\to 1} g_{22}^2(0) = 8/5$ as $N \rightarrow \infty$.

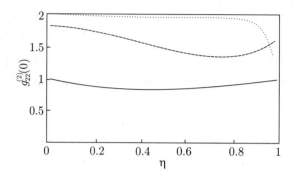

Figure 4. The steady-state dependence of the second-order correlation function against Here solid long-dashed, and short-dashed curves are for $N = 2, 10$ and 300, respectively

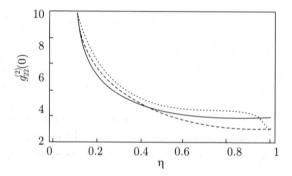

Figure 5. The cross-correlation function against bath parameters Solid, long-dashed, and short-dashed curves are for $N = 2, 10$ and 300, respectively

The cross-correlations $g_{i \neq j}^{(2)}(0)$ also show non-classical behavior. For an atomic sample in a weak bath, $g_{i \neq j}^{(2)}(0)$ is much larger than unity as shown in Figure 5, indicating super-Poissonian light statistics, which is accompanied by strong correlation between the fluorescence light radiated on both atomic transitions, i.e., cross super-bunching. The reason is that then atoms which decay from $|1\rangle \rightarrow |2\rangle$ also decay further to $|3\rangle$ with a high probability. For stronger baths, however, larger samples exhibit bunched sub-Poissonian light. Then

$$\lim_{\{\eta_1, \eta_2\} \rightarrow 1} g_{12}^{(2)}(0) = 4(N + 2)(N + 5)/[5N(N + 3)],$$

with limit $4/5 < 1$ for $N \rightarrow \infty$. In this case, atoms decaying from $|1\rangle$ are $|2\rangle$ re-pumped by the bath rather than decaying further to $|3\rangle$ As an application for the non-classical features, we now show that the light emitted from the sample of atoms violates the Cauchy-Schwarz inequalities (CSI). The CSI are violated if:

$$\chi_{1(2)} = \frac{g_{11}^{(2)}(0) g_{22}^{(2)}(0)}{[g_{12(21)}^{(2)}(0)]^2} < 1,$$

i.e., if the cross correlations between photons emitted on two different transitions are larger than the correlation between photons emitted from the individual levels. Figure 6 shows the violation of the CSI function for moderately strong baths. Within the Dicke model, this violation is present for any number of atoms in the sample, thus demonstrating a macroscopic quantum effect. In addition, χ_1 is always smaller 1 than unity for $N \bullet 3\bullet$ and for the entire range of η, while χ_2 is larger than unity for any number of atoms and for any value of η[19].

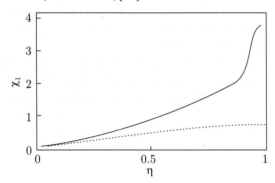

Figure 6. The Cauchy-Schwarz parameters χ_1 shown vs bath parameters $\eta \cdot \eta_1 = \eta_2$ The solid and long-dashed lines are plotted for $N = 3$ and 100, respectively

IV. SUMMARY

In summary, we have demonstrated that a laser-driven mixture of two different two-level ensembles generates highly refractive media without absorption. The group velocity of a weak electromagnetic field pulse probing the laser-driven atomic samples may be abruptly altered and depends sensitively on the external atomic and laser parameters. Second, we have shown that the steady-state quantum dynamics of three-level ladder-type radiators can be influenced by surrounding thermal fields. In particular, large cross- correlations together with a violation of the CSI indicate quantum entanglement among the photons emitted by incoherently driven atoms.

REFERENCES

1. R. H. Dicke, Phys. Rev., 1954, 93, 99.
2. G. S. Agarwal, Quantum Statistical Theories of Spontaneous Emission and Their Relation to Other Approaches, Springer, Berlin, 1974.
3. L. Allen and J. H. Eberly, Optical Resonance and Two-Level Atoms, Wiley, New York, 1975.
4. M. Gross and S. Haroche, Phys. Rep., 1982, 93, 301.
5. A. V. Andreev, V. I. Emel'yanov and Yu. A. Il'inskii, Cooperative Effects in Optics. Super-fluorescence and Phase Transitions, IOP Publishing, London, 1993.
6. R. R. Puri, Mathematical Methods of Quantum Optics, Springer, Berlin, 2001.
7. Z. Ficek and S. Swain, Quantum Interference and Coherence: Theory and Experiments, Springer, Berlin, 2005.

8. C. H. Keitel, M. O. Scully and G. Sussmann, Phys. Rev. A, 1992, 45, 3242.

9. V. Kozlov, O. Kocharovskaya, Y. Rostovtsev and M. O. Scully, Phys. Rev. A, 1999, 60, 1598.

10. F. Haake et al., Phys. Rev. Lett., 1993, 71, 995; Phys. Rev. A, 1996, 54, 1625.

11. M. Macovei, J. Evers, G.-X. Li and C. H. Keitel, unpublished.

12. T. Brandes, Physics Reports, 2005, 408, 315.

13. J. M. Taylor, A. Imamoglu and M. D. Lukin, Phys. Rev. Lett., 2003, 91, 246802.

14. S. Inouye et al., Science, 1999, 285, 571; D. Schneble et al., Science, 2003, 300, 475.

15. M. Macovei, J. Evers and C. H. Keitel, Phys. Rev. Lett., 2003, 91, 233601; Europhys. Lett., 2004, 68, 391; Phys. Rev. A, 2005, 71, 033802.

16. M. Macovei and C. H. Keitel, J. Phys. B: At. Mol. Opt. Phys., 2005, 38, L315.

17. D. F. Walls and G. J. Milburn, Quantum Optics, Springer-Verlag, Berlin, 1994.

18. M. O. Scully and M. S. Zubairy, Quantum Optics, Cambridge University Press, Cambridge, 1997.

19. M. Macovei, J. Evers and C. H. Keitel, Phys. Rev. A., 2005, 72, 06380.

Atoms and Molecules in Laser and External Fields

Editor: Man Mohan

Copyright © 2008, Narosa Publishing House, New Delhi, India

Atoms and Molecules in Strong, High-Frequency Fields

J. P. Hansen, M. Førre, S. Selstø and I. Sundvor

Department of Physics and Technology, University of Bergen, Allégt. 55,
N-5007 Bergen, Norway

INTRODUCTION AND METHOD

Laser technology has gone through continuous improvements ever since the beginning. By now, ultra intense lasers pulses with photon energies as high as 100 eV and with duration on the atto second time scale has been demonstrated [1, 2]. The interaction between such fields and matter is a highly non perturbative one which calls for accurate, ab initio modelling methods on the theoretical side.

The method for solving the Schrödinger equation that we apply, is an extension of the original scheme of Hermann and Fleck [3]. Through a uniformly distributed spherical quadrature made public by Wommersley and Sloan [4], the approach is able to treat the electron dynamics in all three spatial dimensions [5]. The reduced wave function is expanded in Spherical Harmonics,

$$\Theta(\overrightarrow{r}, t) = \sum_{l,m} f_{l,m}(r,t) Y_{l,m}(\hat{r}), \tag{1}$$

and the Schrödinger equation (in atomic units),

$$\left\{ -\frac{1}{2}\frac{\partial^2}{\partial r^2} + \frac{L^2}{2r^2} + V_s(r) + W(\overrightarrow{r}, t) \right\} \Theta(\overrightarrow{r}, t) = i\frac{\partial}{\partial t}\Theta(\bar{r}, t), \tag{2}$$

is solved by writing the time propagator as

$$\Theta(\overrightarrow{r}, t+\Delta t) = e^{-iA\Delta t/2}e^{-iB\Delta t/2}e^{-iC\Delta t}e^{-iB\Delta t/2}e^{-iA\Delta t/2}\Theta(\overrightarrow{r}, t) + O(\Delta t^2),$$

$$A \equiv -\frac{1}{2}\frac{\partial^2}{\partial r^2},$$

$$B \equiv l(l+1)/(2r^2) + V_s(r),$$

$$C \equiv W(\overrightarrow{r}, t) \tag{3}$$

The spatial differentiations are carried out through fast Fourier transforms. While the spherical part of the potential amounts to a straight forward multiplication of the spherical components of the wave function, it is necessary to construct and afterwards decompose the entire wave function to propagate the anisotropic, time-dependent potential W.

In principle any effective one electron dynamic system may be described by this method. In atomic and molecular physics, the most obvious applications would be collisions [6] and interaction with light. In the following we will focus on three examples of the latter.

APPLICATIONS

In general we will take the field to be of the form:

$$\vec{A}(t) = \frac{\vec{E_0}}{\omega} \sin^2 \left(\frac{\pi t}{T} \right) \sin(\omega t + \varphi), \tag{4}$$

where E_0 is the maximum field strength, ω is the central frequency, T is the pulse duration and φ is carrier envelope phase.

In the following we will focus on the dependence of geometry in the interaction between the laser pulse and anisotropic systems. Furthermore, we will study non dipole effects and its importance to atomic stabilisation.

Laser Ionisation of the Hydrogen Molecular Ion

This most simple of molecular systems has been subject to intense theoretical investigation. Still, no ab initio, non perturbative description of this system including all spatial variables has been achieved. This has to do with the systems high number of degrees of freedom. In addition to the three spatial dimensions for the electronic, the internuclear separation R and the relative orientation θ between the internuclear axis and the polarisation of the field need to be considered, making the description far more complex than the corresponding atomic one.

Validity of the Fixed Nuclei Approach

Due to the long rotational period of the molecule it is usually quite safe to assume the orientation of the internuclear axis to be fixed during the time of interaction. In our case, we are focusing on pulses as short as 500 as. Therefore it seems reasonable to assume that also the internuclear separation R may be held fixed. Figure 1 confirms this [7]. It shows the total ionisation probability for a one dimensional electron with fixed internuclear separation, with classical internuclear dynamics and with full quantum mechanical internuclear dynamics.

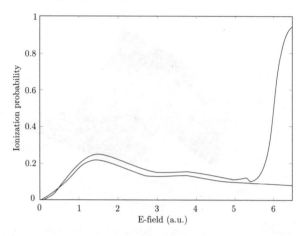

Figure 1. Ionisation probability with fixed nuclei (dotted curve), classical internuclear dynamics (dashed curve) and full quantum mechanical description (full curve). The laser has a central frequency of $\omega = 1$ a.u. (about 27 eV), and the pulse duration T is 2 fs

Results

The Schrodinger equation is solved with the Hamiltonian

$$H_l = -\frac{1}{2}\nabla^2 - \frac{1}{|\vec{r} + \vec{R}/2|} - \frac{1}{|\vec{r} - \vec{R}/2|} + \vec{E}(t) \cdot \vec{r} \tag{5}$$

in the length gauge. The electric field $\mathbf{E(t)}$ defines the angle θ with the internuclear axis. Figure 2 shows the total ionisation probability P_I as a function of both the internuclear distance R and the orientation θ [8].

The ionisation probability exhibits strong dependence on both R and θ. In particular, for $\theta = 0°$ the ionisation probability P_I oscillates with R. For $\theta = 90°$ these oscillations are absent, however. This phenomenon can be understood in terms of interference between outgoing waves originating from each of the scattering centres. Assuming that each of the outgoing waves essentially travels in the direction of the field, two outgoing waves in the direction parallel to the internuclear axis will have an initial phase difference depending on their separation R making the total outgoing wave subject to destructive interference. In the direction perpendicular to the internuclear axis, there is no initial phase difference, and hence no such interference effect either.

Also the angular distribution of the photo electron can be understood within the same idea. Figure 3 shows the probability of ionisation as a function of direction for parallel $(= 0°)$, intermediate $(= 45°)$ and perpendicular $(= 90°)$ polarisation. The first two rows correspond to $R = 2$ and 3, respectively, whereas the third row represents an average over R corresponding to the vibrational ground state.

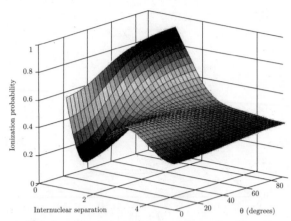

Figure 2. Ionisation probability as a function of internuclear distance R and the angle between the field and the internuclear axis. The laser field is given by $= 2$ a.u., $E_0 = 3$ a.u. and the field duration T corresponds to 6 optical cycles

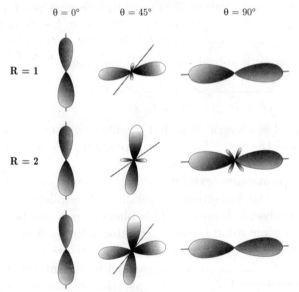

Figure 3. Angular distributions of the photo electron for various orientations and internuclear separations. The lowest row of figures corresponds to an initial R-distribution given by the vibrational ground state. The field parameters are the same as in Figure 2

ORIENTATIONAL DEPENDENCE IN PHOTO-IONISATION OF H(2P)

We may expect that the relative orientation θ will influence the ionisation probability also in the case of non isotropic atomic states. We have investigated this influence for a hydrogen atom initially prepared in the $n = 2, l = 1, m = 0$-state exposed to laser field of energy $\omega = 1$ a.u. [9]. Results are displayed in Figure 4. This time θ refers to the angle between the field and the quantisation axis. We find a rather strong dependence on this angle.

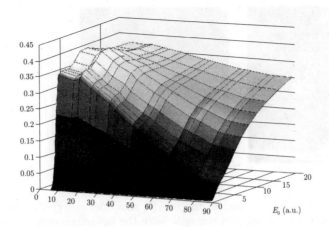

Figure 4. Probability of ionizing $H(2p)$ as a function of the orientation of the electric field (θ) and the maximum field strength (E_0). The central frequency is $\omega = 1$ a.u., and the pulse duration corresponds to 5 optical cycles

It is well known that the multi photon ionisation channels close as the photon energy becomes very large compared to the effective ground state binding energy [10]. For $\theta = 90°$, the $1s$-state is not accessible through a one photon transition $(\Delta m = 1)$ causing the effective binding energy to be lower than in the $\theta = 0°$-case. Consequently, the perpendicular ionisation is suppressed compared to the parallel one. Furthermore, as the intensity of the field increases, the effective binding energy decreases and the multi photon channels close for all values of θ. Due to this phenomena, the ionisation probability ceases to increase with the field intensity, and the system is stabilised. These phenomena explains the behaviour in Figure 4 at higher field intensities.

ATOMIC STABILISATION AND NONE-DIPOLE EFFECTS

The effect of stabilisation has also been studied in more detail for a hydrogen atom initially in the ground state. In the stabilisation limit, the effective ionisation potential is small compared to the photon energy. This is manifested in the fact that only the zeroth order Floquet term in the potential in the Kramers Henneberger formulation of the Hamiltonian [11-13] contributes to the interaction,

$$V_{KH}(\vec{r}, t) = -\frac{1}{|\vec{r} + \vec{\alpha}(t)|} \rightarrow V_{KH}^0(\vec{r}) = -\frac{1}{T} \int_0^T \frac{1}{|\vec{r} + \vec{\alpha}(t)|}. \qquad (6)$$

This potential represents the time average of the field. Thus, the dynamics in this limit arises as a consequence of the non-adiabatic turn on and off of the laser pulse. We expect that as the intensity of the field increases, the energy spectra of the photo electron loose the peaks corresponding integer numbers of the photon energy, and only the strong maximum near the threshold survives. This is exactly what we find, as demonstrated in Figure 5 [14].

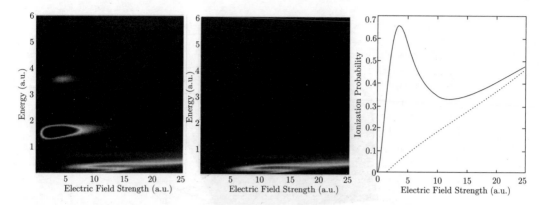

Figure 5. The figure to the left and in the middle show the energy distribution of the ionisation probability for various field strengths with the full interaction with the field and with the time averaged potential, respectively. We see that they coincide as E_0 becomes large. The figure to the right shows the total ionisation probability in the same two cases (full curve and dotted curve, respectively). Here we have $\omega = 2$ a.u. and T corresponds to 5 optical cycles. The shape of the pulse is slightly different from Eq. (4)

It has been claimed that atomic stabilisation is an artefact of the dipole approximation and that the effect will be strongly reduced by inclusion of the magnetic field [15]. This claim can be checked using the non-dipole version of the Kramers Henneberger Hamiltonian:

$$H_{KH}^{ND} = -\frac{1}{2}\nabla^2 + V(+(\eta)) + \frac{1}{2}(A(\eta))^2,$$

$$\eta \equiv \vec{k} \cdot \vec{r} - \omega t. \tag{7}$$

This formula is valid as long as the condition

$$\frac{E_0}{\omega c} \ll 1 \tag{8}$$

where E_0 is given in atomic units, is fulfilled. From Figure 6 we see that as long as E_0 stays below 20 a.u. (corresponding to an intensity of about 10^{19} W/cm^2), the total ionisation probability is essentially unaltered by the inclusion of the magnetic field. The figure shows the ionisation probability and the survival probability of the ground state within and without the dipole approximation. It should be noted, however, that other quantities, such as e.g. the angular distribution of the photo electron, may be influenced by the spatial variation of the field, even though the total ionisation probability is not.

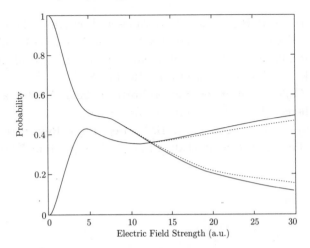

Figure 6. Ionisation probability and survival probability of the ground state with $\omega = 2$ a.u. and $T = 380$ a.s.. The full curve corresponds to the full interaction, and the dashed one to the dipole approximation

CONCLUDING REMARKS

We have demonstrated that for non-isotropic molecular and atomic systems, geometry is crucial, and hence any adequate description of the dynamics should include all three spatial degrees of freedom of the electron. Furthermore, the effect of atomic stabilisation is found to sustain inclusion of the magnetic field of the laser. It has been demonstrated that the phenomena is a consequence of the closing down of multi photon ionisation channels. Consequently, photo electrons are found to have a very low energy after being ionized. The effect of the magnetic field on the photo electron, in particular its angular distribution, is currently being investigated. Furthermore, methods to include more particles in the scheme is being developed.

REFERENCES

1. A. Baltuska, Th. Udem, M. Uiberacker, M. Hentschel, E. Goulielmakis, Ch. Gohle, R. Holzwarth, V. S. Yakovlev, A. Scrinzi and T. H. Hànsch, F. Kraysz, Nature (London), 412, 611 (2003).
2. R. Kienberger, E. Goulielmakis, M. Uiberacker, A. Baltuska, V. Yakovlev, F. Bammer, A. Scrinzi, Th. Westerwalbesloh, U. Kleinberg, U. Heinzmann, M. Drescher and F. Krausz, Nature (London), 427, 817 (2004).
3. M. R. Hermann and J. A. Fleck, Phys. Ref. A, 38, 6000 (1988).
4. I. H. Sloan and R. S. Wommersley, Adv. Comput. Math. (2003),
   http://www.maths.unsw.edu.au/rsw/Sphere
5. J. P. Hansen, T. Matthey and T. Sorevik, A parallel split operator mehtod for the time dependent Schrödinger equation, 10th Euro PVM/MPI (2003).
6. J. P. Hansen, T. Sørevik and L. B. Madsen, Phys. Rev. A, 68, 031401 (R) (2003).
7. I. Sundvor, J. P. Hansen and R. Taïeb, submitted to Phys. Rev. A.

8. S. Selstø, M. Førre, J. P. Hansen and L. B. Madsen, Phys. Rev. Lett., 95, 093002 (2005).

9. T. Birkeland, M. Førre, J. P. Hansen and S. Selstø, J. Phys. B, 37, 4205 (2004).

10. M. Gavrila, Atoms in Intense Laser Fields, edited by Gavrila, Academic, San Diego, 1992.

11. W. Pauli and M. Fierz, Nuovo Cimento, 15, 167 (1938).

12. H. A. Kramers, Collected Scientific Papers, North Holland, Amsterdam, 1956.

13. W. C. Hennebeberger, Phys. Rev. Lett., 21, 838 (1968).

14. M. Førre, S. Selstø, J. P. Hansen and L. B. Madsen, Phys. Rev. Lett., 95, 043601. (2005).

15. A. Bugacov, M. Pont and R. Shakeshaft, Phys. Rev. A, 48, R4027 (1993).

Atoms and Molecules in Laser and External Fields

Editor: Man Mohan

Copyright © 2008, Narosa Publishing House, New Delhi, India

Trapping Population Inside a Potential Well by Using Laser Field

G. Rahali[1,2], R. Taïeb[2], A. Makhoute[1,3] and A. Maquet[2]

[1] UFR de Physique Atomique, Moléculaire et Optique Appliquée, Faculté des Sciences, Université Moulay Ismaïl, B.P. 11201, Zitoune, Meknés, Morocco

[2] Laboratoire de Chimie Physique-Matiére et Rayonnement, Université Pierre et Marie Curie, 11 Rue Pierre et Marie Curie, 75 231 Paris Cedex 05, France

[3] The Abdus Salam International Centre for Theoretical Physics, strada costiera, II − 34100 Trieste, Italy

1. INTRODUCTION

In recent years, the development of novel computational resources such as parallel computers, and numerical techniques such as effcient, high order representations of the derivative operator, have facilitated use of the lattice technique, which is probably, the most direct approach to the solution of the *Time-dependent Schrödinger Equation* (TDSE). With this approach, the evolution of few-body atomic systems is treated directly by solving the TDSE on a spatial lattice. Interest in this approach stems from the potential it holds for producing results that obviate the need for perturbative treatment in regimes where the interactions are strong and that avoid some of the shortcomings associated with other standard methods such as multi center atomic-orbital close coupling or molecular-orbital close coupling. In earlier work this computationally demanding three-dimensional problem was commonly reduced to two dimensions by the assumption of rotational invariance of the wavefunction with respect to the internuclear axis [1]. For non-zero impact parameters this is in general a severe approximation which, however, was weakened by a modest analytical expansion in magnetic substates in [2].

Meanwhile, reasonably converged three-dimentional calculations have become possible due to the availability of more powerful computers. In the approaches of Schurtz *et al.* [3-5], and Horbatsch [6] the discretization is performed on a Cartesian lattice. In order to avoid reflections of the evolving electronic wavefunction at the boundaries of the lattice-box, an absorbing potential is introduced.

The finiteness of the box and the spacing of the grid limits the number of atomic orbitals which can be represented on the lattice typically to states whith principal quantum number $n < 4$. Transition probabilities are calculated by projection of the dynamical wavefunction onto these lattice and total ionization is then obtained by subtracting the bound states probabilities from unity. Schultz et al. estimate that their results for antiproton scattering obtained with 135^3 grid points are accurate to within 10 percent. Achieving higher accuracy implies a re nement of the lattice, which is straihgtforward but computationally expensive.

An appealing feature of the lattice solutions is, that one obtains direct information about the ionization distribution, e.g. by studying the time-propagation of the total density in position space [4] or by transformation of the ejected electron distribution to momentum space [6].

Accordingly, a direct solution of the Schrödinger equation in momentum space has been proposed very recently by Sidky and Lin [7]. It has been pointed out that the problem of reflections at the boundaries does not occur in momentum space since the time-evolving wavefunction can be set to zero at large momenta [8]. In the approach of Sidky and Lin the wavefunction is constructed in momentum space by a two-center expansion in spherical harmonics with radial functions represented by B-splines.

The principal challenge in pursuing this application of the lattice technique is to maintain to the greatest extend possible the characteristics of the approach that make it unique. That is, one seeks to treat the collision directly without recourse to approximations other than those inherent in the numerical representation and propagation of the electronic wave function.

In the present work, we apply this approach to treat capture of electron by using the trapping with laser field in one dimension. In section 2. we present our theoretical treatment and numerical technique. In the end we analyze our results by including a comparison with those obtained for free electron described by a Gaussian wave packet and without laser field. The atomic system (au) of units is used throughout unless otherwise noted.

2. THEORY AND NUMERICAL METHOD

Let us consider a free electron of mass m and charge $- e$ embedded in both scattering potential V and laser field represented by his potential vector $\boldsymbol{A}$. The corresponding non-relativistic Schrödinger equation is given by

$$\left[\frac{1}{2}(\boldsymbol{p} + \boldsymbol{A})^2 + V\right]\psi(r,t) = H\psi(\boldsymbol{r},t) = i\frac{\partial}{\partial t}\psi(r,t) \tag{1}$$

where H is the Hamiltonian of the electron in the presence of scattering potential and laser field, $\boldsymbol{r}$ denotes the space coordinate of the electron and $\boldsymbol{p} = \boldsymbol{k}$ represent the electron wave vector. We shall assume that the laser field is treated classically as spatially homogenous electric field linearly polarized and single mode

$$\mathcal{E}(t) = \mathcal{E}_0 \sin(\omega t + \varphi) \tag{2}$$

with the corresponding vector potential in the Coulomb gauge is

$$A(t) = A_0 \cos(\omega t + \varphi), \tag{3}$$

with $A_0 = \omega \mathcal{E}_0/c$, $\mathcal{E}_0$ and ω are the peak electric field strength and the laser angular frequency, respectively. Here j denotes the initial phase of the laser field.

Without external field and for $V = 0$, equation(1) has the plane-wave solution. A plane wave describes the ideal situation of a particle having a perfectly well defined momentum, but which is *completely delocalized*. Describing a particle with a wave packet enables us to have fairly precise values of both momentum and position.

We consider a free electron initially described by a wave packet, namely

$$\varphi(x,0) = \frac{1}{\sqrt{2\pi}} \int g(k,0) \exp(ikx)dk \tag{4}$$

where $g(k;0)$ is simply the Fourier transform :

$$g(k,0) = \frac{1}{\sqrt{2\pi}} \int \varphi(x,0) \exp(-ikx)dx. \tag{5}$$

In the Gaussian model the wave packet equation(4) is obtained by superposing planes waves $\exp(ikx)$ with the coefficients

$$g(k,0) = \frac{\sqrt{a}}{(2\pi)^{1/4}} \exp\left(-\frac{a^2}{4}(k-k_o)^2\right), \tag{6}$$

which correspond to a Gaussian function centered at $k = k_o$, where a denote the width and ko the group velocity of the Gaussian wave packet. Its center moves uniformly and is located at $x = x_o + k_o t$.

We may imagine that the laser is "switched on" at a certain instant $t = 0$ and "switched off" at the time t while remaining time-independent between the two times 0 and t. Thus we may write in this case the solution of equation (1) as

$$\psi(x,t) = \exp[-iH(t-t_o)]\psi(x,t_o), \tag{7}$$

where the Hamiltonian operator is independent of time. By this we mean that even when the parameter t is changed, the Hamiltonian operator remains unchanged (sudden approximation of laser field).

To resolve the equation we use the Crank-Nicholson numerical method technique which consists in calculating numerically the wave function φ at any moment by discretizing it on a basis, namely

$$x_j = x_o + j\Delta x \qquad j = 0, 1, \ldots, n \tag{8}$$

$$t_m = t_o + m\Delta t \qquad m = 0, 1, \ldots, m_{\max}, \tag{9}$$

where x_o is the position of the wave packet center at the moment $t_o = 0$, Δx and Δt are respectively spacing between two subdivisions on the x-axis and the time axis. For the finite difference representation of $\exp(-iH\Delta t)$ we use the form of Cayley [9]. Thus the solution of Schrodinger equation is given by the system

$$\sum_{j=0}^{n}(1+iH\Delta t/2)_{l_j}\psi_j^{m+1} = \sum_{j=0}^{n}(1-iH\Delta t/2)_{l_j}\psi_j^{m}, \tag{10}$$

where ℓ range from 0 to n and ψ_j^m denote $\psi(x_j, t_m)$ coordinates of $\psi_m(x)$ on the spacial grid.

In matrix representation, the system of equations (10) could be written in the following form as

$$A\psi_{m+1}(x) = B\psi_m(x), \tag{11}$$

where A and B are respectively matrices associated to the operators $(1+iH\Delta t = 2)$ and $(1 - iH\Delta t = 2)$, and $\psi_m(x)$ represent the wave function of the electron at the moment t_m.

3. RESULTS AND DISCUSSION

We resolved numerically the system of equations (11) for, $\Delta x = 0.1$ au, $\Delta t = 0.005$ au, $n = 2000$ and $m_{\max} = 2000$, which enabled us to see the propagation of wave packet of an electron in a box of $l = 200$ au of length, centered on zero for various values of t between 0 and 100.

In order to understand the basic physics we fellow a simple approach which solves the problem approximately and explains all the features observed in the experiment, consideration of one dimension model for a free electron capture is a reasonable first test of the applicability of our technique. In Figures 1-3, we present our results for the density of the presence probability of the electron as a function of the abscise for different cases, namely, free electron, electron near a scattering potential and electron moving in the combined field of the Colombian potential well and of the laser field.

A.   Free Electron

In this case, the diagonal and off diagonal elements of A matrix are, respectively,

$$a_{jj} = 1 + i\frac{\Delta t}{2}\frac{1}{\Delta x^2} \quad \text{and} \quad a_{jj'} = \frac{-i}{4}\frac{\Delta t}{\Delta x^2} \quad \text{with} \quad j^{-1} = j \pm 1 \tag{12}$$

and those of B matrix are

$$b_{jj} = 1 + i\frac{\Delta t}{2}\frac{1}{\Delta x^2} \quad \text{and} \quad b_{jj'} = \frac{i}{4}\frac{\Delta t}{\Delta x^2} \quad \text{with} \quad j^{-1} = j \pm 1 \tag{13}$$

In Figure 1, we present the evolution of the wave packet describing a free electron, we observe clearly that the wave packet get lower by time but also wider such way to preserve its norm constant.

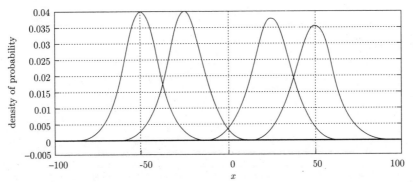

Figure 1. Propagation of a free Gaussian wave packet with ko = 1.5 au and a = 20 au.

B. Electron in a Potential Field

In our calculation to avoid divergence at 0 we considered a regularized Colombian potential [10-12], namely

$$V(x_n) = -\frac{1}{\sqrt{c^2 + x_n^2}} \quad \text{with} \quad c = \sqrt{2}. \tag{14}$$

In the case of a free electron near a potential field, the matrix elements of A and B are respectively

$$a_{jj} = 1 + \frac{i}{2}\frac{\Delta t}{\Delta x^2} + V_j\frac{i\Delta t}{2} \quad \text{and} \quad a_{jj'} = \frac{-i}{4}\frac{\Delta t}{\Delta x^2} \quad \text{with} \quad j^{-1} = j \pm 1 \tag{15}$$

and those of B are

$$b_{jj} = 1 + \frac{i}{2}\frac{\Delta t}{\Delta x^2} - V_j\frac{i\Delta t}{2} \quad \text{and} \quad b_{jj'} = \frac{i}{4}\frac{\Delta t}{\Delta x^2} \quad \text{with} \quad j^{-1} = j \pm 1 \tag{16}$$

with $V_j = V(x_j)$ is the value of the potential at the position x_j.

In Figure 2 we show the effect of the potential on the propagation of the wave packet of the electron, we note that it accelerates after its crossing of the potential well and its speed increases by one atomic unit.

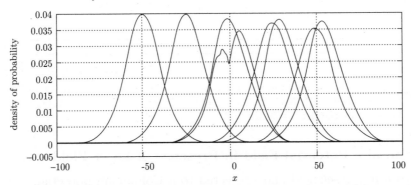

Figure 2. Propagation of the wave packet of an electron in absence of potential (curve with lines) and in presence of regularized Colombian potential $V(x) = -(x^2 + 2)^{-1/2}$ (curve with dots). $k_0 = 1$ au, $a = 20$ au and $x_0 = -50$ au

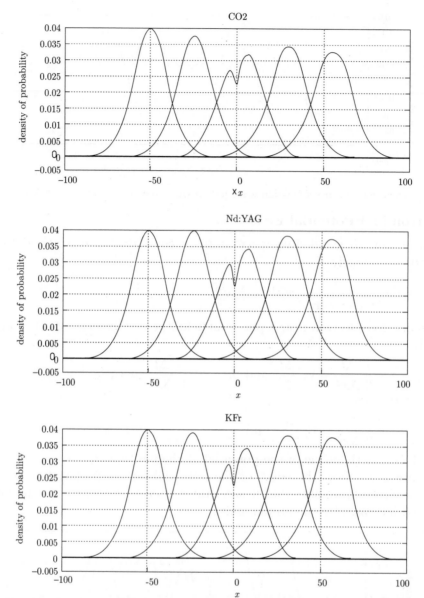

Figure 3. Propagation of a Gaussian wave packet trough a Colombian potential for various laser fields: CO_2 ($\omega = 0.0043$ au); Nd : Y AG ($\omega = 0:043$ au) and KFr ($\omega = 0.184$au), with $k_0 = 1$ au, $a = 20$ au and $x_0 = -50$ au

C.   Electron in a Potential Field in Presence of Laser Field

For an electron in a laser field and subjected to a Colombian potential. The elements of the matrix A and B in this case are

$$a_{jj} = 1 + i\frac{\Delta t}{2}\left(\frac{1}{\Delta x^2} + \frac{A^2}{2} + V_j\right) \text{ and } a_{jj'} = \frac{A}{4}\frac{\Delta t}{\Delta x}\cos\theta - \frac{i}{4}\frac{\Delta t}{\Delta x^2} \text{ with } j^{-1} = j\pm 1 \quad (17)$$

and those of B are

$$b_{jj} = 1 - i\frac{\Delta t}{2}\left(\frac{1}{\Delta x^2} + \frac{A^2}{2} + V_j\right) \text{ and } b_{jj'} = -\frac{A}{4}\frac{\Delta t}{\Delta x}\cos\theta + \frac{i}{4}\frac{\Delta t}{\Delta x^2} \text{ with } j^{-1} = j\pm1 \quad (18)$$

with θ is the angle between the momentum p and the potential vector A.

In set of Figure 3 we present the computation results for various laser frequencies. We noted that when the frequency decreases, the number of electrons trapped inside the potential well increases. We observe a striking decrease in the state density on the outlet side of the well when the frequency gets 10 times less. The numerical method used in this work enabled us to solve the equation of one-dimensional *Time-Dependant Schrodinger Equation* (TDSE), for an electron crossing a Colombian potential well in a medium where a laser field reigns. We observed the trapping phenomena of electrons by representing the density of the presence probability of the electron in the space in which it moves since it penetrates inside the well until that it crosses it to the other side. As perspectives of this work, we are aiming to study the evolution of the wave packet in the case of two and three dimensions in view to study, by after, the interaction electron-atom in the presence of a laser field. The atom, in our study, enters in interaction not only with just one electron but rather with an electrons beam of which speeds undergo fluctuations. We think of describing the incident trajectories in a classical statistical way by calling upon the method of Monte-Carlo. Then we will investigate the effect of the mentioned fluctuations, as well as the fluctuations of laser parameters, on the electron - atom interaction mechanisms.

REFERENCES

1. V. Maruhn-Rezwani, N. Grün and W. Scheid, Phys. Rev. Lett., 43 (1979), 512; C. Bottcher, ibid, 48 (1982), 85; K. C. Kulander, K. R. S. Devi and S. E. Koonin, Phys. Rev. A, 25 (1982), 2968.
2. N. Grün, A. Mühlhans and W. Scheid, J. Phys. B, 15 (1982), 4043.
3. D. R. Schultz, P. S. Krstic, C. O. Reinhold and J. C. Wells, Phys. Rev. Lett., 76 (1996), 2882.
4. J. C. Wells, D. R. Schultz, P. Gavras and M. S. Pindzola, Phys. Rev. A, 54 (1996), 593.
5. D. R. Schultz, J. C. Wells, P. S. Krstic and C. O. Reinhold, Phys. Rev. A, 56 (1997), 3710.
6. M. Horbatsch, Proc. 14th Int. Conf. on the Application of Accelerators in Research and Industry, Denton, TX, American Institute of Physics, 32 (1997), 71.
7. E. Y. Sidky and C. D. Lin, J. Phys. B, 31 (1998), 2949.
8. K. Momberger, A. Belcakem and A. H. Sorensen, Phys. Rev. A, 53 (1996), 1605.
9. W. H. Press, B. P. Flannery, S. A. Teukolsky and W. T. Vetterling, Numerical Recipes The Art of Scientific Computing (FORTRAN Version), 637, 640–642, Cambridge University Press, New York, 1989.
10. C. Bottcher and M. R. Strayer, Ann. Phys., 175 (1987), 64.
11. J. Javanainen, J. H. Eberly and Q. Su, Phys. Rev. A, 38 (1988), 3430.
12. Q. Su and J. H. Eberly, Phys. Rev. A, 44 (1991), 5997.

Atoms and Molecules in Laser and External Fields

Editor: Man Mohan

Copyright © 2008, Narosa Publishing House, New Delhi, India

Photodissociation Dynamics:
Polarization of Atomic Fragments

Alex Brown[1], G.G. Balint-Kurti[2] and O.S. Vasyutinskii[3]

[1] Department of Chemistry, University of Alberta, Edmonton, Alberta, T6G 2G2, Canada

[2] School of Chemistry, University of Bristol, Bristol BS8 1TS, United Kingdom

[3] Ioffe Physico-Technical Institute, Russian Academy of Sciences,194021 St. Petersburg, Russia

1. INTRODUCTION

The hydrogen halides, HX (X = F, Cl, Br and I), provide excellent model systems for examining photodissociation dynamics involving multiple excited electronic states. Experimental and theoretical studies, for example, see Refs. [1-15], have focused on the photodissociation process:

$$HX + hn \rightarrow H(^2S) + X(^2P_{3/2}) \tag{1a}$$
$$\rightarrow H(^2S) + X(^2P_{1/2}) \tag{1b}$$

producing ground state, $X(^2P_{3/2})$, and spin-orbit excited state, $X(^2P_{1/2})$, halogen atoms. The energy differences between the two product channels are 404 cm^{-1}, 882 cm^{-1}, 3685 cm^{-1}, and 7603 cm^{-1}, for HF, HCl, HBr, and HI, respectively. Therefore, the effects of increasing spin-orbit coupling on the dissociation process can be examined systematically. An understanding has been sought of the roles of the various electronic states involved in the excitation and the possible non-adiabatic transitions that could take place between the excited potential energy curves (PECs) as the molecule fragments. The total cross-section, the branching fraction Γ, which quantifies the yield of spin-orbit excited atoms, $X(^2P_{1/2})$, relative to the total yield, and the anisotropy parameter β, which provides information on the parallel and/or perpendicular nature of the electronic transitions contributing to the dissociation, have been the primary subjects of experimental and theoretical studies. While the measurement or computation of these properties has provided a wealth of information on the potential energy curves, transition dipole moments, and, if applicable, non-adiabatic couplings, underlying the dynamics, our recent work [1, 4, 5, 8, 11, 12] has focused on calculating the $a_Q^{(K)}(p)$ parameters describing the orientation and alignment of the halogen atoms in the molecular frame. The halogen atoms have angular momenta: $j_X = 3/2$ and $1/2$ for ground state and spin-orbit excited state halogen atoms, respectively. Therefore, they can have a preferred orientation and/or alignment in space and this can be fully described in the molecular frame by the $a_Q^{(K)}(p)$ parameters [16]. K and Q refer to the spatial distributions in the molecular frame. The symmetry of the transition dipole moments from the ground electronic state to the dissociating states is given by p and can be $\parallel$, $\perp$, or $(\parallel, \perp)$ corresponding to pure parallel, pure perpendicular, or mixed parallel/perpendicular excitation. Our theoretical work has, in part, been motivated by the beautiful experiments of Rakitzis et al. [5, 17, 18] who have measured the alignment and orientation of chlorine and bromine atoms resulting from the dissociation of HCl and HBr, respectively.

The paper is organized as follows. In Section 2, the theoretical and computational techniques needed for determining the $a_Q^{(K)}(p)$ parameters are briefly reviewed. In Section 3, we discuss the results for a single anisotropy parameter $a_0^{(2)}(\perp)$ for the four hydrogen halides and highlight the new insight that has been obtained into the dissociation dynamics. In Section 4, we summarize the results and suggest possible future research directions.

2. THEORY

The first step required in a theoretical investigation of photodissociation is to determine the underlying potential energy curves, the corresponding electronic transition dipole moments, and, if required, the non-adiabatic couplings between the electronic states. We have used electronic structure data obtained from ab initio calculations for HF [2], HCl [7], HBr [8], and HI [14]. If rotation is neglected ($J = 0$), there are eight electronic states involved in the photodissociation of the hydrogen halides: the ground state and seven excited states optically accessible via one-photon transitions. Five of these states, $X^1\Sigma_{0+}$ (non-degenerate ground state), $A^1\Pi_1$ (doubly degenerate), and $a^3\Pi_1$ (doubly degenerate), correlate with the production of ground state halogen atoms, see Eq. (1a). Three states, $a^3\Pi_{0+}$ (non-degenerate) and $t^3\Sigma_1$ (doubly degenerate), correlate with the production of excited state halogen fragments, see Eq. (1b). The term symbols represent a mixed Hund's case (a)/case (c) according to $^{2S+1}L_\Omega$. When including the spin-orbit coupling, Ω is the only good quantum number and the ^{2S+1}L labels designate the largest case (a) contribution within the Franck-Condon region. For the light hydrogen halides, HF and HCl, the initial excitation is dominated ($> 99.5\%$) by the $A^1\Pi_1 \leftarrow X^1\Sigma_{0+}$ transition. Therefore, any production of excited state halogen atoms arises due to non-adiabatic coupling of the $\Omega = 1$ states. On the other hand, for HBr and HI, where the spin-orbit mixing of the electronic states is significant, there can be direct excitation to all optically allowed states.

Once the electronic structure is known, a time-dependent wave packet treatment [2, 6] is utilized to examine the non-adiabatic quantum dynamics; for HI, the dynamics was assumed to be adiabatic [11, 12]. From the dynamics, a set of energy dependent coefficients, $A_n(R_\infty, t)$, can be determined:

$$A_n(R_\infty, E) = \frac{1}{2\pi} \int_0^\infty \phi_n(R_\infty, t) e^{i(E_i + h\nu)/h} dt. \tag{2}$$

where $\phi_n(R_x, t)$ is the time-dependent wave function for electronic state n evaluated at a large fixed value of the bond length $(R = R_x)$. R_x is chosen such that there is no longer coupling between the adiabatic electronic states. From these energy dependent coefficients, the photofragmentation T matrix element associated with each channel n can be calculated:

$$\langle \Psi_n^- \Omega(R, E) | \hat{d}_q | \Psi_\Omega \rangle = i \left(\frac{h^2 k_v}{2\pi\mu} \right)^{\frac{1}{2}} e^{(-ik, R_\infty)} A_n(R_\infty, E). \tag{3}$$

where $\hat{d}_q$ is the transition dipole moment operator, k_v is the asymptotic wave vector for this dissociation channel, μ is the reduced mass of the photofragments, and $\Psi_{n\Omega}^-(R, E)$ is the scattering wave function for channel (n, Ω) in the body-fixed coordinate system.

The dimensionless anisotropy parameters $a_Q^{(k)}(p)$, which fully define the angular momentum distribution of the photofragments in the molecular frame, are normalized combinations of the dynamical functions, $f_K(q, q')$. These parameters can be determined via a well-established theoretical framework [16, 20-22]. The relationships between the dynamical functions and the $a_Q^{(k)}(p)$ parameters are described in Ref. [1]. The dynamical functions for the halogen atoms resulting from the dissociation of HX leading to photofragments with angular momenta j_X and j_H depend on the photofragment T matrix elements and are given by

$$f_K(q, q') = \sum_{\substack{n, \Omega, \Omega_X, \\ n'\Omega'\Omega_X'}} (-1)^{K + j_X + \Omega_X'} \begin{pmatrix} j_X & j_X & K \\ -\Omega_X & \Omega_X' & q - q' \end{pmatrix} (T_{j_X^\Omega X_j H^\Omega H}^{n\Omega})^* T_{j_X \Omega' X_j H^\Omega H}^{n'\Omega'}$$

$$\times \langle \Psi_{n,\Omega}^-(R, E) | \hat{d}_q | \Psi_{\Omega_i} \rangle^* \times \langle \Psi_{n',\Omega'}^-(R, E) | \hat{d}_q' | \Psi_{\Omega_i} \rangle \tag{4}$$

The $T_{j_X^\Omega X_j H^\Omega H}^{n\Omega}$ are expansion coefficients of the adiabatic electronic molecular wave functions at large internuclear distances in terms of the wave functions of the separate atoms, see Ref. [1]. The indices q and q' are the vector spherical harmonic components of the molecular electric dipole moment with respect to the recoil axis and can take the values 0 or ± 1 corresponding to parallel or perpendicular electronic transitions, respectively. The initial and final z-components of the total electronic angular momentum about the molecular axis are related by $\Omega = \Omega_1 + q$. The diagonal $(q = q')$ and off-diagonal $(q \neq q')$ elements of the dynamical functions $f_K(q, q')$ correspond to incoherent and coherent excitation of different molecular continua. The dynamical functions relevant for a photofragment with angular momentum j_X range from $K = 0$ to $K = 2j_X$, where K is referred to as the multipole rank.

In our studies of the hydrogen halides [1, 4, 8, 11], the $a_0^{(K)}(\perp)$ parameters ($K = 1, 2$, and 3) describing incoherent perpendicular excitation, the $a_2^{(K)}(\perp)$ parameters ($K = 2, 3$) describing coherent perpendicular excitation, and the $a_1^{(1)}(\parallel, \perp)$ parameter describing coherent parallel and perpendicular excitation, have been determined for the ground state fragments $X(^2P_{3/2})$. For the excited state fragments $X(^2P_{1/2})$, the $a_0^{(1)}(\perp)$ and $a_1^{(1)}(\|, \perp)$ parameters have been computed. For illustrative purposes, only the incoherent anisotropy parameter $a_0^{(2)}$ describing the production of ground state halogen atoms is discussed here. The $a_0^{(2)}(\perp)$ parameter is related to the dynamical functions by

$$a_0^{(2)}(\perp) = \left[\frac{(2j_X + 3)(2j_x - 1)}{j_X(j_x + 1)} \right]^{1/2} \frac{f_2(1,1)}{f_0(1,1)} = \frac{4}{\sqrt{5}} \frac{f_2(1,1)}{f_0(1,1)}. \tag{5}$$

The incoherent parameter $a_0^{(2)}(\perp)$ is of interest because it can be related to the amount of ground state atoms $X(^2P_{3/2})$ produced via the $a^3\Pi_1$ electronic state versus that produced from both the $a^3\Pi_1$ and $A^1\Pi_1$ states, i.e.,

$$a_0^{(2)} = \frac{2(1 - 2p)}{5} = \frac{4}{5} - \frac{8}{5} \left(\frac{\sigma_a^3\Pi_1}{\sigma_{a^3\Pi_1} + \sigma_{A^1\Pi_1}} \right) \tag{6}$$

where σ_n is the partial cross-section for state n. Information regarding p cannot be obtained by measurement of the branching fraction since both electronic states lead to $X(^2P_{3/2})$ products or the lowest order anisotropy parameter β as both states are accessed via perpendicular excitation. There are two limiting cases for the value of $a_0^{(2)}(\perp)$: if 100% of the halogen atoms are produced via $A^1\Pi_1$, it will equal 0.8 while if 100% of the halogen atoms are produced via $a^3\Pi_1$, it will equal -0.8.

3. RESULTS AND DISCUSSION

3.1. HF and HCl

For HF and HCl, the experimental measurement or theoretical determination of $a_0^{(2)}(\perp)$ provides firsthand information on the non-adiabatic coupling from the $A^1\Pi_1$ electronic state to the $a^3\Pi_1$ state since only the $A^1\Pi_1$ state is populated via direct excitation from the ground state. Figure 1 illustrates the $a_0^{(2)}(\perp)$ anisotropy parameter describing incoherent perpendicular excitation for the ground state $X(^2P_{3/2})$ fragment produced from the photodissociation of HF and HCl in their ground vibrational states, $(v = 0)$, as a function of photolysis wavelength. Our predictions agree well with the single experimental measurement at 193 nm for HCl [17].

The results presented here are exact, within the limitations of the electronic structure calculations upon which they are based, but it is useful to consider them within the context of three simple static models: the adiabatic model, the diabatic or sudden model, and the statistical model. In the adiabatic model, it is assumed that the fragments part so slowly that the system remains on a single electronic state. For HF and HCl, since direct excitation takes place to $A^1\Pi_1$ only, the adiabatic model would predict all of the products to be those correlating with this electronic state. Within the adiabatic model, $a_0^{(2)} = 0.8$. In the diabatic model, it is assumed that (i) the atoms separate so quickly that there is no time for the electronic structure to follow the adiabatic correlation diagram and (ii) the rearranging of all molecular quantum states happens simultaneously within a narrow window of the internuclear distance. The diabatic model predicts that the halogen atoms are produced one-half in $A^1\Pi_1$ and one-sixth in $a^3\Pi_1$ leading to $X(^2P_{3/2})$ atoms and, with one-third produced in $t^3\Sigma_1$ giving excited state atoms. These distributions give $a_0^{(2)}(\perp) = 0.4$. Finally, in the statistical model, all accessible states are equally populated, which results in $a_0^{(2)}(\perp) = 0$. Clearly, none of these static models reproduce the exact results presented in Figure 1, including the strong dependence on photolysis energy. For HCl, we have investigated a simple dynamical model based on the two-state exponential one, but this has met with limited success in reproducing the full quantum dynamical results [4]. It will be extremely interesting if further experimental measurements can be made as a function of photolysis wavelength as has been done for other properties such as the excited state branching fraction for HCl.

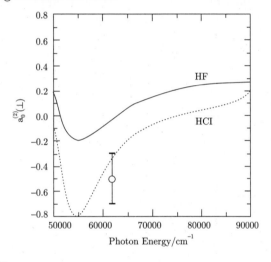

Figure 1. Incoherent $a_0^{(2)}(\perp)$ anisotropy parameter for the production of $X(^2P_{3/2})$ as a function of photon energy for the photodissociation of HF (solid line) and HCl (dashed line). In order to present results on the same scale, the energy scale has been shifted by $+10000$ cm^{-1} for the HCl data. Also, shown is the experimental measurement for HCl at 193 nm [17]

3.2. HBR and HI

For HBr and HI, direct excitation to all optically accessible excited states occurs and, therefore, the products arise from both direct excitation (adiabatic) and non-adiabatic coupling mechanisms. All available ab initio and empirical models for HI assume that the dissociation proceeds adiabatically, and adiabatic dynamics calculations based on the best available ab initio data [14] have successfully reproduced experimental measurements of the total cross-section, branching fraction, and β parameter. For HBr, the electronic structure data including the non-adiabatic couplings is available [8] but it is interesting to compare exact non-adiabatic quantum dynamics results to fully adiabatic ones where the non-adiabatic coupling is "turned off". Figure 2 illustrates the $a_0^{(2)}(\perp)$ anisotropy parameter describing incoherent perpendicular excitation for the ground state $X(^2P_{3/2})$ fragment produced from the photodissociation of HBr and HI in their ground vibrational states, $(v = 0)$, as a function of photolysis wavelength. Also, shown is $a_0^{(2)}(\perp)$ predicted for HBr within the adiabatic model. Our exact predictions, including the non-adiabatic couplings, agree well with the single experimental measurement at 193 nm for HBr [17]. The non-adiabatic transitions play a significant role in HBr dissociation as the adiabatic prediction at 193 nm is significantly different from the experimental measurement.

For the adiabatic HBr and HI models, Figure 2 shows the dominance of the absorption to the $a^3\Pi_1$ state at low energy $(a_0^{(2)}(\perp) = -0.8)$ switching to absorption to the $A^1\Pi_1$ state at high energy $(a_0^{(2)}(\perp) = 0.8)$. The switch simply reflects the relative energies of these two electronic states. As expected, the results for neither system agree with the predictions of the statistical model $(a_0^{(2)}(\perp) = 0)$. We are currently investigating the predictions of the diabatic model for HBr. Until experimental measurements are made for HI either confirming or conflicting with the theoretical predictions [11], we will assume that the dynamics is purely adiabatic for HI.

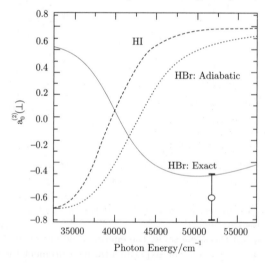

Figure 2. Incoherent $a_0^{(2)}(\perp)$ anisotropy parameter for the production of $X(^2P_{3/2})$ as a function of photon energy for the photodissociation of HBr including non-adiabatic coupling (solid line), HBr within the adiabatic model (dashed line) and HI (solid line). Also, shown is the experimental measurement for HBr at 193 nm [17]

4. CONCLUSIONS

For the hydrogen halides (HF, HCl, HBr, and HI), the alignment and orientation of the halogen atoms resulting from photodissociation have been shown [1,4,8,11] to be extremely sensitive to the details of the excited states, their non-adiabatic couplings, and the transition dipole moments. More importantly, they provide more detailed information than measurements of the total cross-section, the excited state branching fraction, or the lowest-order anisotropy parameter β. As an example, exact quantum dynamical predictions of the incoherent parameter $a_2^{(2)}(\perp)$ for ground state atoms $X(^2P_{3/2})$ have been presented and discussed in the context of simple models: the adiabatic, the diabatic, and the statistical. None of the simple models can fully explain the exact quantum dynamical results. The exact theoretical predictions are in quantitative agreement with the experimental measurements at 193 nm for HCl and HBr. Hopefully, our theoretical predictions will encourage further experimental measurements as a function of photolysis wavelength for which theory predicts large photon-energy dependent effects. Additionally, experimental measurement of $a_0^{(2)}(\perp)$ for HI should be able to answer once and for all whether its dissociation is truly adiabatic.

Acknowledgments

The authors thank the many people with whom they have had the privilege to collaborate with on various aspects of this work. Our collaborators include T.P. Rakitzis, J.A. Beswick, A.J. Orr-Ewing. A.G. Smolin and D.N. Jodoin. AB thanks the Natural Sciences and Engineering Research Council of Canada and the University of Alberta for financial support.

REFERENCES

1. G.G. Balint-Kurti, A.J. Orr-Ewing, J.A. Beswick, A. Brown and O.S. Vasyutinskii, Vector correlations and alignment parameters in the photodissociation of HF and DF, J. Chem. Phys., 116(2002), 10760–10771. .

2. A. Brown and G.G. Balint-Kurti, Spin-orbit branching in the photodissociation of HF and DF: I. A time-dependent wavepacket study for excitation from $v = 0$, J. Chem. Phys., (2000), 1870–1878.

3. J. Zhang, C.W. Riehn, M. Dulligan and C. Wittig, An experimental study of HF photodissociation: Spin-orbit branching ratio and infrared alignment, J. Chem. Phys., 104 (1996), 7027–7035.

4. A. Brown, G.G. Balint-Kurti and O.S. Vasyutinskii, Photodissociation of HCl and DCl: Polarization of atomic photofragments, J. Phys. Chem. A, 108 (2004), 7790–7800.

5. T.P. Rakitzis, P.C. Samartzis, R.L. Toomes, T.N. Kitsopoulos, A. Brown, G.G. Balint-Kurti, O.S. Vasyutinskii and J.A. Beswick, Spin-polarized hydrogen atoms from molecular photodissociation, Science, 300 (2003), 1936-1938.

6. P.M. Regan, D. Ascenzi, A. Brown, and G.G. Balint-Kurti and A.J. Orr-Ewing, Ultraviolet photodissociation of HCl in selected rovibrational states: Experiment and theory, J. Chem. Phys., 112 (2000), 10259–10268.

7. M.H. Alexander, B. Pouilly and T. Duhoo, Spin-orbit branching in the photofragmentation of HCl, J. Chem. Phys., 99 (1993), 1752–1764.

8. A.G. Smolin, O.S. Vasyutinskii, G.G. Balint-Kurti and A. Brown, Photodissociation of HBr. 1. Electronic structure, photodissociation dynamics, and vector correlation coefficients, J. Phys. Chem. A (in press), DOI: 10.1021/jp0562429

9. P.M. Regan, S.R. Langford, M.N.R. Ashfold and A.J. Orr-Ewing, The ultraviolet photodissociation dynamics of hydrogen bromide, J. Chem. Phys., 110 (1999), 281–288.

10. B. Pouilly and M. Monnerville, New investigation of the photodissociation of the HBr molecule: total cross-section, anisotropy parameter and dependence of the spin-orbit branching on the ground state vibrational level, Chem. Phys., 238 (1998), 437–444 .

11. D.N. Jodoin and A. Brown, Photodissociation of HI and DI: testing models for electronic structure via polarization of atomic photofragments, J. Chem. Phys., 123 (2005), 054301.

12. A. Brown, Photodissociation of HI and DI: polarization of atomic photofragments, J. Chem. Phys., 122 (2005), 084301.

13. J.P. Camden, H.A. Bechtel, D.J.A. Brown, A.E. Pomerantz, R.N. Zare and R.J. LeRoy, Probing excited electronic states using vibrationally mediated photolysis: application to hydrogen iodide, J. Phys. Chem. A, 108 (2004), 7806–7813.

14. A.B. Alekseyev, H.P. Liebermann, D.B. Kokh and R.J. Buenker, On the ultraviolet photofragmentation of hydrogen iodide, J. Chem. Phys., 113 (2000), 6174–6185.

15. P. M. Regan, D. Ascenzi, C. Clementi, M.N.R. Ashfold and A.J. Orr-Ewing, The UV photodissociation of HI revisited: REMPI measurements of I(2P) atom spin-orbit branching fractions, Chem. Phys. Lett., 315 (1999), 187–193.

16. T. P. Rakitzis and R. N. Zare, Photofragment angular distributions in the molecular frame: Determination and interpretation, J. Chem. Phys., 110 (1999), 3341–3350.

17. T. P. Rakitzis, P. C. Samartzis, R. L. Toomes, L. Tsigaridas, M. Coriou, D. Chestakov, A. T. J. B. Eppink, D. H. Parker and T. N. Kitsopoulos, Photofragment alignment from the photodissociation of HCl and HBr, Chem. Phys. Lett., 364 (2002), 115-120.

18. T. P. Rakitzis, P. C. Samartzis, R. L. Toomes and T. N. Kitsopoulos, Measurement of Br photofragment orientation and alignment from HBr photodissociation: production of highly spin-polarized hydrogen atoms, J. Chem. Phys., 121 (2004), 7222–7227.

19. B. V. Picheyev, A. G. Smolin and O. S. Vasyutinskii, Ground state polarized photofragments study by using resonance and off-resonance probe beam techniques, J. Phys. Chem. A, 101 (1997), 7614–7626.

20. A. S. Bracker, E. R. Wouters, A. G. Suits and O. S. Vasyutinskii, Imaging the alignment angular distribution: State symmetries, coherence effects, and nonadiabatic interactions in photodissociation, J. Chem. Phys., 110 (1999), 6749–6765.

21. L. D. A. Siebbeles, M. Glass-Maujean, O. S. Vasyutinskii, J. A. Beswick and O. Roncero, Vector properties in photodissociation: quantum treatment of the correlation between the spatial anisotropy and the angular momentum polarization of the fragments, J. Chem. Phys., 100 (1994), 3610–3623.

22. E. R. Wouters, M. Ahmed, D. S. Peterska, A. S. Bracker, A. G. Suits and O. S. Vasyutinskii, Imaging the atomic orientation and alignment in photodissociation, in Imaging in Chemical Dynamics, edited by A. G. Suits and R. E. Continetti, American Chemical Society, Washington DC, 2000, pp. 238–284.

Atoms and Molecules in Laser and External Fields

Editor: Man Mohan

Copyright © 2008, Narosa Publishing House, New Delhi, India

Selective Control of Photodissociation in HOD

M. Sarma[1], S. Adhikari[1,2], S. Deshpande[1], Vandana K.[1] and Manoj K. Mishra[1]

[1] Department of Chemistry, Indian Institute of Technology, Powai, Mumbai, India
[2] Department of Chemistry, Indian Institute of Technology, Guwahati, North Guwahati, India

1. INTRODUCTION

Attempts to use lasers as molecular scissors to cleave bonds selectively are being pursued extensively (Akagi *et al.*, 2005; Crim, 1993; Gordon and Rice, 1997; Rabitz *et al.*, 2000; Rice and Zhao, 2000; Shapiro and Brumer, 2003; Zare, 1998). The established theoretical (Brumer and Shapiro, 1992; Gross *et al.*, 1991; Judson and Rabitz, 1992; Shi *et al.*, 1988; Tannor and Rice, 1985) and experimental schemes (Akagi *et al.*, 2005; Assion *et al.*, 1998; Baumert and Gerber, 1994; Crim, 1990; Cohen *et al.*, 1995; Lu *et al.*, 1992; Vander Wal *et al.*, 1991) rely on designing appropriate laser pulses to achieve the desired outcome from photodissociation reactions, often requiring field attributes that cannot be foreseen on the basis of chemical considerations (Baumert and Gerber, 1994; Gross *et al.*, 1992) and may also be difficult to reproduce (Gross *et al.*, 1994) in normal laboratory conditions.

The deuterated water molecule HOD has been a popular prototype for investigation of selective control through vibrational mediation (Akagi *et al.*, 2005; Amstrup and Henriksen, 1992; Cohen *et al.*, 1995; Crim, 1990; Lu *et al.*, 1992; Vander Wal *et al.*, 1991). The H-OD (3706 cm-1) and HO-D (2727 cm^{-1}) stretching frequencies are well separated. This provides for selective excitation of more or less pure $O-H$ and $O-D$ modes and the use of either selectively excited higher $O-H$ overtones (Crim, 1993) or a combination of IR and UV pulses to first produce large quanta of vibrational stretching in the desired bond and then pump it to the lowest repulsive excited electronic state for selective photodissociation, has been investigated in detail in many groups (Akagi *et al.*, 2005; Amstrup and Henriksen, 1992; Cohen *et al.*, 1995; Crim, 1990; Imre and Zhang, 1989; Lu *et al.*, 1992; Elghobashi *et al.*, 2003; Vander Wal *et al.*, 1991; Zhang *et al.*, 1989). Amstrup and Henriksen (Amstrup and Henriksen, 1992) have theoretically investigated both active and passive approaches to selective control of HOD photodissociation except that the UV fields employed were δ-function type or with extremely narrow 5 fs pulse width and also required large quanta of vibration ($n_{O-D} = 4$) in the O − D stretch. The pulse utilized by Elghobashi *et al.* (2003) are similarly narrow in time domain with a width of ~ 4000 cm^{-1} making it difficult to decipher vibrational mechanism behind preferential cleavage of O − H&O − D bonds in HOD.

It is our purpose in this paper to present some results which show that considerable selectivity and yield in HOD dissociation may be achieved with easily realizable small quanta vibrational excitations and more realistic UV fields. Furthermore, we have been advocating the use of Field Optimized Initial State (FOIST) scheme (Gross *et al.*, 1996; Vandana and Mishra, 1999a, 1999b, 2000) which attempts to distribute the onus for selective control on both the field attributes and the molecular initial state to be subjected to the chosen photolysis pulse both of which can be sampled separately and economically using simple Time Dependent Quantum Mechanical (TDQM) techniques based on Fast Fourier Transform (FFT) (Kosloff and Kosloff, 1983) and Lanczos propagation (Leforestier *et al.*, 1991) to select a combination of field attributes which may be easier to realize experimentally. Earlier applications of the FOIST scheme to HI and IBr photodissociations have provided new insights and encouraging selectivity and yield (Vandana and Mishra, 2000) and this paper provides some preliminary results from our applications to selective cleaving of O − H and O − D bonds in the HOD molecule. Results from application of the FOIST scheme to HOD molecule are analyzed to try and garner features which may facilitate easier routes to enhanced selectivity and yield.

The systemic and methodological details are presented in the following section. In section 3 we discuss our results and a brief summary of principal observations in section 4 concludes this paper.

2. METHOD

Photofragmentation of the deuterated water molecule HOD in the first absorption band involving electronic transition from the ground electronic state $(\widetilde{X}^1A_1)$ to the first excited electronic state $(\widetilde{A}^1B_1)$ takes place on the first excited potential energy surface (Figures 1(a) and 1(b)). The lowest energy excitation induces negligible change in the bending angle, and since the bending is not active in first absorption band (Amstrup and Henriksen, 1992; Imre and Zhang, 1989), the internal kinetic energy operator in terms of the conjugate momenta $\hat{p}_1$ and $\hat{p}_2$ associated with the O − H(r_1) and O − D(r_2) stretching coordinates respectively, is given by

$$\hat{T} = \frac{\hat{p}_1^2}{2\mu_1} + \frac{\hat{p}_2^2}{2\mu_2} + \frac{\hat{p}_1\hat{p}_2}{m_O}\cos_\theta \,, \tag{1}$$

where $\hat{p}_J = \frac{\hbar}{i}\frac{\partial}{\partial r_i}$, $j = 1, 2$, $\mu_1 = \frac{m_H m_O}{m_H + m_O}$, $\mu_2 = \frac{m_O m_D}{m_O + m_D}$, and is the fixed bending angle, 104.52 deg.

The first and second excited electronic states are well separated and following earlier investigation (Amstrup and Henriksen, 1992) we too formulate the HOD dynamics considering only the ground and the first excited electronic states (Figures 1(a) and 1(b)) of the molecule. The time evolution of the corresponding nuclear motion can then be performed using the time dependent Schrdinger equation,

$$ i\hbar \frac{\partial}{\partial t} \begin{pmatrix} \Psi_g \\ \Psi_c \end{pmatrix} = \begin{pmatrix} \hat{H}_g & \hat{H}_{uv}(t) \\ \hat{H}_{uv}(t) & \hat{H}_e \end{pmatrix} \begin{pmatrix} \Psi_g \\ \Psi_c \end{pmatrix} \tag{2} $$

where $\Psi_g = \Psi_g(r_1, r_2, t)$ and $\Psi_e = \Psi_e(r_1, r_2, t)$ are the wave functions associated with nuclear motion in the ground and first excited state, respectively. $\hat{H}_g = \hat{T} + \hat{V}_g$ and $\hat{H}_e = \hat{T} + \hat{V}_e$ are the nuclear Hamiltonians for the two electronic states where $\hat{H}_{uv}(t) = \mu_{ge}(r_1, r_2) E_0 \cos_{uv} t$ (Amstrup and Henriksen, 1992) couples as well as perturbs both the electronic states. We solve Eqn. (2) with the $(t = 0)$ initial condition that the ground state wave function Ψ_g is a single, field free, vibrational state or a linear combination of more than one vibrational state(s) of the HOD electronic ground state, and the excited wave function $\Psi_e = 0$, at $t = 0$.

Vibrational eigenfunctions of the ground electronic state of the HOD molecule were obtained using the Fourier Grid Hamiltonian (FGH) method (Marston and Balint-Kurti, 1989) modified for two dimensions (Dutta *et al.*, 1993). As can be seen from Figures 1(c)–1(f), the vibrational eigenfunction have clear characterization as $|m, n\rangle$ eigenmodes with m and n quanta of excitation in the O $-$ H and O $-$ D bonds respectively. The nodal patterns and eigenfrequencies match those listed by Amstrup and Henriksen and utilized elsewhere by us (Sarma *et al.*, 2006).

In the FOIST scheme (Gross *et al.*, 1994; Vandana and Mishra, 1999a, 2000), the product yield is maximized through preparation of the initial wavefunction $|\Psi_g(r_1, r_2)\rangle$ as a superposition of the field free vibrational wavefunctions $\{\Psi_m\}$ of the ground electronic state,

$$ \Psi_g(0) = \sum_{m=0}^{M} C_m \Psi_m \tag{3} $$

The product yield in the desired channel is related to the time integrated flux f given by,

$$ \begin{aligned} f &= \int_0^T dt \langle \Psi(t) | \hat{j} | \Psi(t) \rangle \\ &= \int dt \langle \Psi_g(0) | \hat{U}^\dagger(t, 0) \hat{j} \hat{U}(t, 0) | \Psi_g(0) \rangle \\ &= \langle \Psi_g(0) | \hat{F} | \Psi_g(0) \rangle \end{aligned} $$

with

$$ \Psi(t) = \hat{U}(t, 0) \Psi(0), \quad \hat{F} = \int_0^T dt \hat{U}^\dagger(t, 0) \hat{j} \hat{U}(t, 0), $$

and

$$ \hat{j}_1 = \frac{1}{2\mu_1} [\hat{p}_1 \delta(r_i - r_i^d) + \delta(r_i, r_i^d) \hat{p}_i] \tag{4} $$

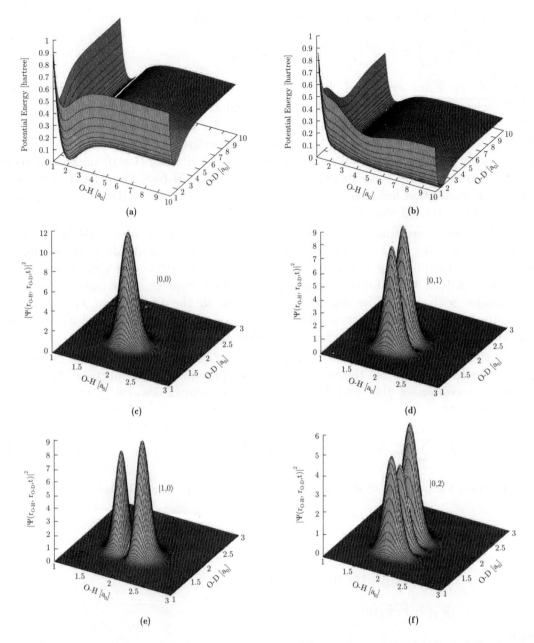

Figure 1. Potential Energy Surfaces for (a) Ground and (b) First Excited of HOD. Probability Density
plots for the $|n_{OH} \cdot n_{OD}\rangle$ modes of HOD are depicted in: (c) $|0.0\rangle$, (d) $|0.1\rangle$, (e) $|1.0\rangle$ and
(f)— $0.2\rangle$

where U is the time evoltution operator, $\hat{j}_i$ is the flux operator in the i-th channel
and $\mu_i, \hat{p}_i$ and r_i^d are the reduced mass, the momentum operator and a grid point
in the asymptotic region of the i-th channel denoted by reaction coordinate r_i, with

$H + O - D$ channel labeled as 1 and $H - O + D$ channel as 2. The expressions for the total flux J in the $H + O - D$ and $H - O + D$ channels are given by,

$$J_{H+O-D} = \int_0^{r_2 d} \int_o^T \Psi^*(r_1, r_2, t) \times \left(\hat{j}_1 + \frac{\mu_1 \cos \theta}{m_o} \right) \Psi(r_1, r_2, t) dr_2, dt \,, \qquad (5)$$

$$J_{H-O+D} = \int_0^{r_1 d} \int_o^T \Psi^*(r_1, r_2, t) \times \left(\hat{j}_2 + \frac{\mu_2 \cos \theta}{m_o} \right) \Psi(r_1, r_2, t) dr_1, dt \qquad (6)$$

where the second operator in Eqns. 5 and 6 represents the effect of kinetic coupling between the $O - H$ and $O - D$ modes. The field dependence of $H(\vec{r}_1, \vec{r}_2, t)$ manifests itself through $U(t, 0) \cong e^{-iHt/\hbar}$ where $H = H_{molecule} + H_{UV}$. Optimization of the channel and field specific flux functional $\langle \Psi(0)|\hat{F}|\Psi(0) \rangle$ with respect to the coefficients C_m employed in Eqn. 3 leads to the Rayleigh-Ritz eigenvalue problem (Vandana and Mishra, 1999b)

$$FC = fC \qquad (7)$$

where the f is the diagonal matrix of the time integrated flux matrix F. The matrix elements of F in the i-th channel are given by (Vandana and Mishra, 2000)

$$F_{kl}^i \approx \Delta t \sum_{n=0}^{N_t} \langle \Psi_k(n\Delta t)|J_i|\Psi_1(n\Delta t) \rangle \qquad (8)$$

where $J_1 = \left(\hat{j}_1 + \frac{\mu_2 \cos \theta}{m_o} \hat{j}_2 \right)$ and $J_2 = \left(\hat{j}_2 + \frac{\mu_1 \cos \theta}{m_o} \hat{j}_1 \right)$.

The propagation and other numerical details are identical to those presented elsewhere (Sarma *et al.*, 2006) by us.

3. RESULTS AND DISCUSSION

FOIST based approach attempts to utilize simple field profiles where yield enhancement can be achieved either through mixing of other initial states (Vandana and Mishra, 1999a, 1999b)or mixing of some other colors which can induce coherent transition to the same final state or mixing of both initial states and lasers of appropriate frequencies (Vandana and Mishra, 2000). This requires that the frequency spectrum of fields utilized be sufficiently narrow to fit our chemically motivated mechanistic models based on excitation from and dumping to the vibrational levels of the ground state. The temporal and frequency profiles of 5 fs field (Amstrup and Henriksen, 1992) utilized by Amstrup and Henriksen is depicted in Figure 2(a) and due to its wide frequency span does not serve our requirements. The simple Gaussian pulse used by us and its power spectrum is depicted in Figure 2(b).

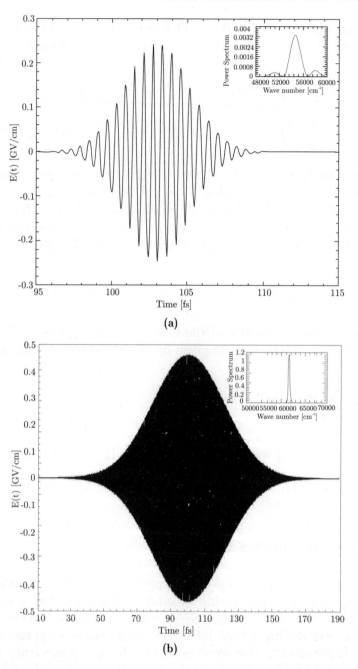

(a)

(b)

Figure 2. The UV field plots for (a) The 5 fs field used by Amstrup and Henriksen (1992) with $\varepsilon(t) = E_0^* \exp(-\gamma(t - t_{uv})^2)^* \cos \omega t : \gamma = 4 ln2 (FWHM)^{-2}$; $FWHM = 5$ fs; $t_{uv} = 103$ fs; $\omega = 54869$ cm^{-1} and maximum field intensity $= 50$ TW/cm^2; (b) The field form employed by us with $FWHM = 50$ fs; $t_{uv} = 100$ fs; $\omega = 60777$ cm^{-1} and maximum field intensity $= 178$ TW/cm^2. The power spectra are shown as insets

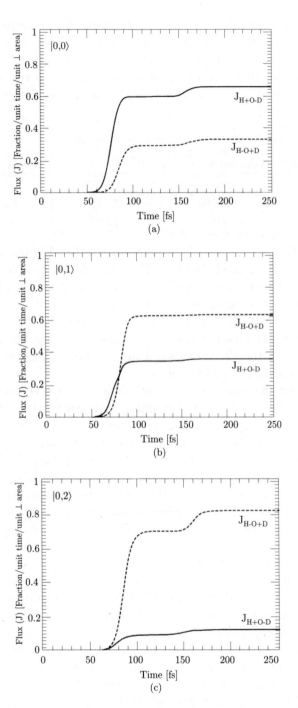

Figure 3. Plot of Flux Vs. Time for (*a*) $|0,0\rangle$, (*b*) $|0,1\rangle$ and (*c*) $|0,2\rangle$

The role of prior stretching of the $O - D$ bond for selective control of $O - D$ photodissociation is seen from flux plots of Figure 3. The accumulated flux in the $H + O - D$ and $H - O + D$ channels with $|0,0\rangle$ as the initial state is plotted in Figure 3(a) which shows that lighter H atom flows faster in the $H + O - D$ channel leading to much larger flux in the $H + O - D$ channel as compared to $H - O + D$ channel (1st row, Table 1).

Table 1. Flux obtained using different initial states and frequencies

Initial state	Frequency (cm^{-1})	$H + O - D$ flux	$H - O + D$ flux
$|0,0\rangle$	60777	65.5%	32.9%
$|0,1\rangle$	59703	35.5%	62.5%
$|0,2\rangle$	54372	12.2%	82.8%

The kinematic factor is also seen to manifest itself in Figure 3(a) as delayed flow of flux in the $H - O + D$ channel. With $|0,1\rangle$ as the initial state, there is greater build up of flux in the $H - O + D$ channel and the reversal of the natural kinematics favoring greater flux in $H + O - D$ channel is reversed quite early and substantively. The final flux in the $H - O + D$ channel (Figure 3(b) and 2nd row of Table 1) is twice as large as that in the $H + O - D$ channel and this reversal of kinematic bias with just one quantum of excitation in the $O - D$ mode using a simple Gaussian pulse offers a fresh alternative for experimental actualization.

Use of $|0,2\rangle$ with two quanta of vibration in the $O - D$ bond (Figure 1(f)) as the initial state and the same Gaussian pulse with carrier frequency corresponding to transition from the $|0,2\rangle$ initial state to the excited state PES (54372 cm^{-1}) provides a surge of probability flow in the $H - O + D$ channel from very beginning and an overwhelming reversal of the favored dissociation pattern in $H + O - D$ channel (Figure 3(c) and 3rd row of Table 1 and Figure 4). Dominant selective dissociation in the $H - O + D$ channel may therefore be obtained with just two quanta of excitation in the $O - D$ bond, using an easily reproducible Gaussian pulse. Both these features should motivate simple experiments for selective photodynamic control of the deuterated water molecule HOD.

Results from FOIST based mixing of additional color and the $|0,0\rangle$ and $|0,1\rangle$ vibrational states with single and two color laser setups are reported in Table 2. Mixing of an additional color is intended to provide field induced dumping from the excited PES to the $|0,1\rangle$ vibrational level of ground PES and thereby facilitate a mixing of $|0,0\rangle$ and $|0,1\rangle$ so that the kinematic bias in favor of $O - H$ dissociation is decreased. Using the $|0,0\rangle$ and $|0,1\rangle$ initial states and a single color photolysis pulse (rows 1 and 2 of Table 2) doesn't show any improvement in the $H - O + D$ flux values obtained with single initial states reported in Table 1. Results from rows 3 and 4 in Table 2 however show that FOIST based mixing of $|0,0\rangle$ and $|0,1\rangle$ using two color fields are quite sensitive to frequencies employed and provide extra means for manipulation of $H + O - D/H - O + D$ ratio. Use of a two color field (54920 cm^{-1} and 52303 cm^{-1}) with $|0,0\rangle$ as initial state shows reversal of natural kinematic bias favoring more flux in the $H + O - D$ channel and provides another avenue for preferential dissociation of the $O - D$ bond.

4. CONCLUDING REMARKS

Selectivity and yield may be influenced by mixing initial states or color in the UV field or both and further attempts to use these variations on HOD are still going on. Preliminary results presented here however provide fresh impetus for utilizing simple, chemically motivated field forms and ideas to achieve considerable selectivity and very high yield in the kinematically unfavorable $H - O + D$ channel using simple initial states like $|0, 1\rangle$ and $|0, 2\rangle$ which should be easy to populate using normal spectroscopic tools.

The probability density profiles reveal richness in the dynamics which with more detailed analysis may provide additional insights for selective control of bond dissociation in HOD and other polyatomics. Effects from change of field width, frequency, amplitude and time/phase delay between multicolor pulses can provide further insights. An effort along these lines is underway in our group.

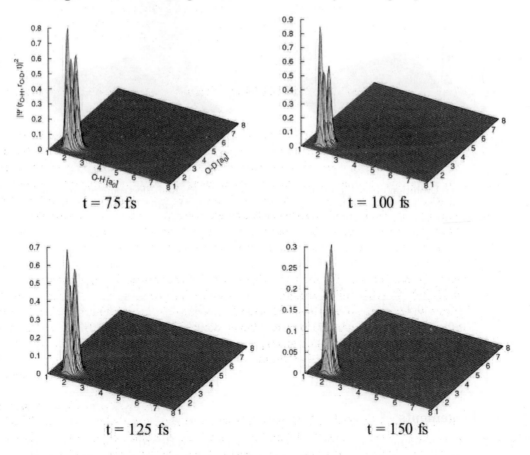

Contd. Figure 4

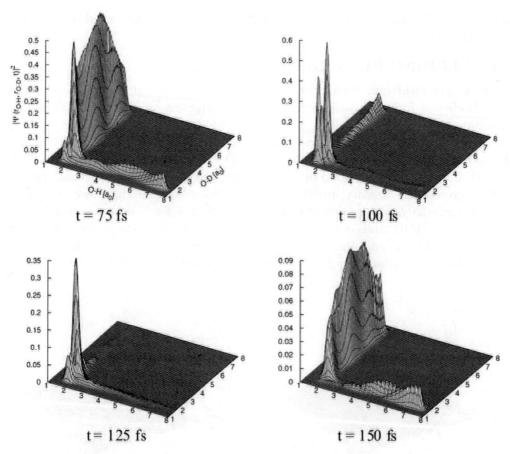

Figure 4. Time Evolution of on the (a) Ground and (b) First Excited Electronic State using the UV field depicted in 2b with carrier frequency $\omega = 54372$ cm^{-1}, corresponding to resonant transition from $|0,2\rangle$ level to the first excited electronic state

Table 2. Flux obtained using different initial states and frequencies

Initial State (s)	Frequencies (cm^{-1})	H + O − D Flux	H − O + D Flux
$|0,0\rangle + |0,1\rangle$	59703	35.5%	65.5%
$|0,0\rangle + |0,1\rangle$	60777	35.9%	59.9%
$|0,0\rangle + |0,1\rangle$	60777 and 59703	31.8%	66.2%
$|0,0\rangle + |0,1\rangle$	60277 and 59203	30.3%	67.9%
$|0,0\rangle$	54920 and 52303	18.5%	52.2%

Acknowledgements

Manoj K. Mishra acknowledges financial support from the Board of Research in Nuclear Sciences (Grant No. 2001/37/8/BRNS) of the Department of Atomic Energy, India. S. Adhikari acknowledges Department of Science and Technology (DST), Government of India for partial financial support through project No. SP/S1/H-53/01. M. Sarma acknowledges support from CSIR, India (SRF, F. No. 9/87(336)/2003-EMR-I).

REFERENCES

1. H. Akagi, H. Fukazawa, K. Yokoyama and A. Yokoyama (2005), Selective $O - D$ bond dissociation of HOD: photodissociation of vibrationally excited HOD in the 5 state, J. Chem. Phys., 123 (18), 184305-1 — 184305-7.

2. B. Amstrup and N. E. Henriksen (1992), Control of HOD photodissociation dynamics via bond selective infrared multiphoton excitation and a femtosecond ultraviolet laser pulse, J. Chem. Phys., 97 (11), 8285–8295.

3. A. Assion, T. Baumert, M. Bergt, T. Brixner, B. Kiefer, V. Seyfried, M. Strehle and G. Gerber (1998), Control of chemical reactions by feedback-optimized phase-shaped femtosecond laser pulses, Science, 282, 919–922.

4. T. Baumert and G. Gerber (1994), Fundamental interactions of molecules (Na2, Na3) with intense femtosecond laser pulses, Isr. J. Chem., 34 (1), 103–114.

5. P. Brumer and M. Shapiro (1992), Laser control of molecular processes, Annu. Rev. Phys. Chem., 43, 257–282.

6. Y. Cohen, I. Bar and S. Rosenwaks (1995), Photodissociation of HOD ($\nu_{OD} = 3$): Demonstration of preferential O-D bond breaking, J. Chem. Phys., 102 (9), 3612-3616.

7. F. F. Crim (1990), State and bond selected unimolecular reactions, Science, 249, 1387–1392.

8. F. F. Crim (1993), Vibrationally mediated photodissociation: Exploring excited state surfaces and controlling decomposition pathways, Annu. Rev. Phys. Chem., 44, 397–428.

9. P. Dutta, S. Adhikari and S. P. Bhattacharyya (1993), Fourier Grid Hamiltonian method for bound states of multidimensional systems:formulations and preliminary applications to model systems, Chem. Phys. Lett., 212 (6), 677–684.

10. N. Elghobashi, P. Krause, J. Manz and M. Oppel (2003), $IR + UV$ laser pulse control of momenta directed to specific products: quantum model simulations for $HOD^* \rightarrow H + OD$ versus $HO + D$, Phys. Chem. Chem. Phys., 5, 4806–4813.

11. R. J. Gordon and S. A. Rice (1997), Active control of the dynamics of atoms and molecules, Annu. Rev. Phys. Chem., 48, 601–641.

12. P. Gross, D. Neuhauser and H. Rabitz (1991), Optimal control of unimolecular in the collisional regime, J. Chem. Phys., 94 (2), 1158–1166.

13. P. Gross, D. Neuhauser and H. Rabitz (1992), Optimal control of curve-crossing systems, J. Chem. Phys., 96 (4), 2834–2845.

14. P. Gross, D. P. Bairagi, M. K. Mishra and H. Rabitz (1994), Optimal control of IBr curve crossing reactions, Chem. Phys. Lett., 223, 263–268.

15. P. Gross, A. K. Gupta, D. P. Bairagi and M. K. Mishra (1996), Initial state laser control of curve crossing reactions using the Rayleigh-Ritz variational procedure, J. Chem. Phys., 104 (18), 7045–7051.

16. D. G. Imre and J. Zhang (1989), Dynamics and selective bond breaking in photodissociation, Chem. Phys., 139, 89–121.

17. R. S. Judson and H. Rabitz (1992), Teaching laser to control molecules, Phys. Rev. Lett., 68 (10), 1500–1503.

18. D. Kosloff and R. Kosloff (1983), A Fourier method solution for the time dependent Schrdinger equation as a tool in molecular dynamics, J. Comp. Phys., 52, 35–53.

19. C. Leforestier, R. H. Bisseling, C. Cerjan, M. D. Feit, R. Friesner, A. Guldberg, A. Hammerich, G. Jolicard, W. Karralein, H. D. Meyer, M. Lipkin, O. Roncero and R. Kosloff (1991), A comparison of different propagation schemes for the time dependent Schrdinger equation, J. Comp. Phys., 94, 59–80.

20. S. P. Lu, S. M. Park, Y. Xie and R. J. Gordon (1992), Coherent laser control of bound-to-bound transitions of HCl and CO, J. Chem. Phys., 96 (9), 661 6620.

21. C. C. Marston, G. G. Balint and Kurti (1989), The Fourier grid Hamiltonian method for bound state eigenvalues and eigenfunctions, J. Chem. Phys., 91 (6), 3571–3576.

22. H. Rabitz, R. D. V. Riedle, M. Motzkus and K. Kompa (2000), Whither the future of controlling quantum phenomena?, Science, 288, 824–828.

23. S. A. Rice and M. Zhao (2000), Optical control of molecular dynamics, Wiley Interscience, New York.

24. M. Sarma, S. Adhikari and M. K. Mishra (2006), Selective control of HOD photodissociation using low quanta O-D excitation and field optimized initial state (FOIST) based combination of states and colors, Chem. Phys. Lett., 420, 321–329.

25. M. Shapiro and P. Brumer (2003), Coherent contol of molecular dynamics, Rep. Prog. Phys., 66, 859–942.

26. S. Shi, A. Woody and H. Rabitz (1988), Optical control of selective vibrational excitation in harmonic linear chain molecules, J. Chem. Phys., 88 (11), 6870–6883.

27. D. J. Tannor and S. A. Rice (1985), Control of selectivity of chemical reaction via control of wave packet evolution, J. Chem. Phys., 83 (10), 5013–5018.

28. K. Vandana and M. K. Mishra (1999a), Photodynamic control using field optimized initial state: A mechanistic investigation of selective control with application to IBr and HI photodissociation, J. Chem. Phys., 110 (11), 5140–5148.

29. K. Vandana and M. K. Mishra (1999b), Selective photodynamic control of chemical reactions: a Rayleigh-Ritz variational approach, Advances in Quant. Chem., 35, 261–281.

30. K. Vandana and M. K. Mishra (2000), A simplification of selective control using field optimized initial state with application to HI and IBr photodissociation, J. Chem. Phys., 113 (6), 2336–2342.

31. R. L. Vander Wal, J. L. Scott, F. F. Crim, K. Weide and R. Schinke (1991), An experimental and theoretical study of the bond selected photodissociation of HOD, J. Chem. Phys., 94 (5), 3548–3555.

32. R. N. Zare (1998), Laser control of chemical reactions, Science, 279, 1875–1878.

33. J. Zhang, D. G. Imre and J. H. Frederick (1989), HOD spectroscopy and photodissociation dynamics: selectivity in OH/OD bond breaking, J. Phys. Chem., 93 (5), 1840–1851.

Atoms and Molecules in Laser and External Fields
Editor: Man Mohan
Copyright © 2008, Narosa Publishing House, New Delhi, India

Spin Dependent Currents in Intense-Field Single Photoionization

S. Bhattacharyya[1,2], Mahua Mukherjee[1], J. Chakraborty[1] and F. H. M. Faisal[2]

[1] Indian Association for Cultivation of Science, Jadavpur, Kolkata, India
[2] Fakultt fr Physik, Universitt Bielefeld, Postfach 100131, D-33501 Bielefeld, Germany

1. INTRODUCTION

Single-photoionization and multi-photoionization in strong fields has been investigated within the framework of Dirac theory in the past but mainly for the spin-unresolved currents (e.g., [1,2]). In this paper we report on the result of analysis of spin-flip probabilities and up-spin (u) and down-spin (d) currents from ionization of an ensemble of Dirac H-atoms subjected to intense *circularly polarized laser* (CPL) radiation. An intensity and frequency-dependent asymmetry between the spin up and down currents that varies according to the direction of electron emission is found. Explicit analytical formulas are derived, and results of numerical calculations are presented using the ground-state Dirac-Volkov wave function with no spin-orbit interaction in the final state [3]. In the position of detection of the ionized electron there is no electromagnetic field, eventually the final electron is taken as a plane wave. The spin-orbit interaction in the ground state is identically zero. With CPL at relativistic intensities, the *angular differential* (AD) rates for $d \rightarrow d$ (dd current) is higher than those from $u \rightarrow u$ (uu current). The spin-flip electron current with $u \rightarrow d$ (ud-current), however small, exists, but the current with $d \rightarrow u$ (du-current) is null. Thus the theory predicts spin-flip du-current as forbidden in the case of single ionization. This is true even when the retardation effect, hence also the magnetic component of the field as well as the spin-orbit interaction responsible for the well-known Fano-effect are negligible. Transformation properties of the up and down spin ionization amplitudes show that the sign of spin can be controlled by a change of helicity of the laser photons from outside [3].

From a straightforward relativistic generalization (e.g., [1]) of the intense-field S-matrix method [4,5], the leading term of the resulting S-matrix series for the transition amplitude for ionization from ground state hydrogen atom $\Psi_{1s}^{(s)}(\vec{r}, t)$ of spin s [5] into continuum states $\Psi_{\vec{p}}^{s'}(\vec{r}, t)$ of spin s' [6], where $(s, s') = u$ (up) or d (down), is given by

$$S_{s \to s'} = \int_{-\infty}^{\infty} \langle \Psi_{\vec{p}}^{s'}(t) | \gamma^{\mu} A_{\mu} | \Psi_{1s}^{(s)}(t) \rangle dt \tag{1}$$

The Dirac-Volkov solution of hydrogen atom

$$\hat{\Psi}_{1s}(\vec{r}, t)^{\uparrow} e^{iE_b t} \Psi_{1s}(\vec{r})^{\uparrow} \tag{2}$$

$$\Psi_{1s}(\vec{r})^{\uparrow} = N_{1s} r^{\gamma'-1} e^{-k_B^r} \begin{pmatrix} 1 \\ 0 \\ i\beta' \cos\theta \\ i\beta' \sin\theta e^{i\phi} \end{pmatrix} \tag{3}$$

$$= N_{1s} r^{\gamma'-1} e^{ik_B^r} \hbar\omega^{\uparrow} = R_{1s}(r) \hbar\omega^{\uparrow} \tag{4}$$

$$\hbar\gamma.n = \gamma_0 n_0 - \overline{\gamma n}, n_{\mu} = (n_0, i\beta'\hat{r}) \tag{5}$$

$$n_0 = 1, \vec{n} = (i\beta'\hat{r})$$

$$\left. \begin{array}{l} k_B = \dfrac{mcz\alpha}{\hbar c} = z \because \alpha = \dfrac{1}{c}, c = 137 \text{in a.u.,} \\[2mm] m = \hbar = |e| = a_0 = 1 \\[4mm] N_{1s} = (2k_B)^{(\chi'+1/2)} \left(\dfrac{1+\gamma'}{8\pi\Gamma(1+2\gamma')} \right)^{1/2} \gamma = \sqrt{1-(Z\alpha)} \\[4mm] \beta' = (1-\gamma')/Z\alpha \end{array} \right\} \tag{6}$$

$$\omega^{\uparrow} = \begin{pmatrix} 1 \\ 0 \\ 0 \\ 0 \end{pmatrix}, \quad \omega^{\downarrow} = \begin{pmatrix} 0 \\ 0 \\ 0 \\ 1 \end{pmatrix} \tag{7}$$

continuum states $\Psi_{\vec{p}}^{(s')}(\vec{r}, t)$ of spin s'

$$\Psi_{\vec{p}}^{(s')}(\vec{r}, t) = N_{p_0} u'_s e^{-irp} \tag{8}$$

$$N_{p_0} = \sqrt{\frac{c}{p_0}},$$

$$u_s = \begin{pmatrix} m_i \chi_s \\ m_2 \sigma \cdot \hat{p} \chi_s \end{pmatrix} \tag{9}$$

up and down spinors are $\chi^{\uparrow} = \begin{pmatrix} 1 \\ 0 \end{pmatrix}, \chi^{\downarrow} = \begin{pmatrix} 0 \\ 1 \end{pmatrix}$.

$$m_1 = \sqrt{\frac{p_0 + c}{2c}}, \qquad m_2 = \sqrt{\frac{p_0 - c}{2c}}$$

For the present purpose we shall restrict the calculations below to the case of a right circularly polarized electromagnetic field ($\xi = +\pi/2$). We choose the field propagation direction (z-axis) as the quantization axis, with the spin "up" state defined to be along the positive z direction. Eventually $\vec{k} = (0, 0, \kappa), \vec{p} = (p_x, p_y, p_z)$.

$$\hat{\varepsilon}(\xi) = \frac{1}{\sqrt{2}}(\hat{\varepsilon}_x, i\hat{\varepsilon}_y, 0) = \sqrt{1}\sqrt{2}(1, i, 0).$$

The electromagnetic field vector

$$\vec{A} = \frac{A_0}{c}[\hat{\varepsilon}_x(\xi)\cos(\omega t - \vec{\kappa}.\vec{r}) - \vec{\varepsilon}_y(\xi)\sin(\omega t - \vec{\kappa}.\vec{r})] \tag{10}$$
$$= \frac{A_0}{2c}(\hat{\varepsilon}(\xi)e^{ikr} + \hat{\varepsilon}^*(\xi)e^{-ikr}),$$

After necessary substitution the S-matrix (1) becomes

$$S_{s \to s'} = -2\pi i \delta(\varepsilon_b + \varepsilon_{kin} - \omega)T_{s \to s'} \tag{11}$$

where $\varepsilon_b = c(c - \sqrt{s^2 - pb^2})$ is the binding energy (ionization potential) and $\varepsilon_{kin} = c(\sqrt{c^2 - P^2} - C)$ is the kinetic energy.

The spin specific reduced $t_{S \to s'}$ matrix is

$$T_{s \to s'} = \frac{A_0 N_{p_0}}{2c} \int R_{1s}(r)\exp[-i(\vec{p} - \vec{k}).\vec{r}](\bar{u}_{s'} \not{\varepsilon}^* \not{h}\omega_s)d^3r$$
$$= \frac{N_{p_0} N_{1s} e A_0}{2c} 4\pi C_0(q)M_{s \to s'} \tag{12}$$

with

$$M_{s \to s'} = \bar{u}_{s'} \not{\varepsilon}^* \not{V}(q)\omega_s, \tag{13}$$

Where

$$V(q) = (1, g(q)\hat{q}), \vec{q} = \vec{p} - \vec{k} = q\hat{q}$$
$$C_0(q) = \frac{4\pi}{q} \frac{\Gamma(\gamma' + 1)}{(p_b^2 + q^2)^{\frac{\gamma'+1}{2}}} \sin\left((\gamma' + 1)\tan^{-1}\left(\frac{q}{p_b}\right)\right) \tag{14}$$

$$g(q) = \beta'\left[\frac{k_B}{q} - \frac{\gamma' + 1}{\gamma'}\sqrt{1 + \left(\frac{k_B}{q}\right)^2} \frac{\sin\left(\gamma'\tan^{-1}\left(\frac{q}{k_B}\right)\right)}{\sin\left((\gamma' + 1)\tan^{-1}\left(\frac{q}{k_B}\right)\right)}\right] \tag{15}$$

$p_0 = \sqrt{c^2 + \vec{p}^2} = \kappa_0 + \sqrt{c^2 - p_b^2}$, the ground state momentum $p_b = Z, \kappa_0 = \omega/c$. After substituting from (9) and (10)

$$M_{s \to s} = \chi_{s'}^+(V - i\vec{\sigma} \cdot \vec{W})\chi_s \tag{16}$$

where $\vec{\sigma}$ is the Pauli matrix, $V = m_1 P - m_2 \hat{p} \cdot \vec{R}$, $\vec{W} = m_1 \vec{Q} + m_2 \hat{p} \times \vec{R}$,

$$P = a \cdot b \cdot \vec{Q} = \vec{a} \times \vec{b}, \ \vec{R} = b_0\vec{a} - a_0\vec{b}, \ a = \varepsilon^*, \ b = V(q).$$

The spin specific matrices (16) are

$$M_{u \to u} = \sqrt{2} m_2 \sin \theta e^{-i\phi} \tag{17}$$

$$M_{u \to d} = \sqrt{2} m_1 \frac{g(q)}{|\vec{q}|} (|\vec{p}| \cos \theta - |\vec{k}|) - m_2 \sqrt{2} \cos \theta \tag{18}$$

$$M_{d \to u} = 0 \tag{19}$$

$$M_{d \to d} = \sqrt{2} m_1 \frac{g(q)}{|\vec{q}|} |p'| \sin \theta e^{-i\phi}. \tag{20}$$

Explicit analytical expressions of the spin-specific probabilities of ionization per unit time, $\frac{dW_{s \to s'}}{d\Omega}$, in the direction (θ, φ) in an element of solid angle $d\Omega = \sin \theta d\theta d\phi$, have been derived [7] by evaluating Eq. (1),

The spin-specific ionization rates of interest are

$$\frac{dW_{s \to s'}}{d\Omega} = \left[\frac{A_0}{2c} N_{p_0} N_{1s} c_0(q) \right]^2 |M_{s \to s'}|^2 \frac{c p_0 |\vec{p}|}{(2\pi)^2} \tag{21}$$

The usual spin unresolved ionization rate for an unpolarized target atom is easily obtained by simply adding the four spin-specific rates and dividing by 2 (for the average with respect to two degenerate initial spin states)

$$\frac{d\Gamma}{d\Omega} = \frac{1}{2} \left[\frac{A_0}{2c} N_{p_0} N_{1s} c_0(q) \right]^2 \sum_{(s,s')=u,d} |M_{s \to s'}|^2 \frac{c p_0 |\vec{p}|}{(2\pi)^2} \tag{22}$$

The spin-up (up) and the spin-down (dn) electron currents can now be obtained from Eqs. (17)-(20) as

$$\frac{dW^{\text{up}}}{d\Omega} = \frac{1}{2} \left(\frac{dW_{u \to u}}{d\Omega} + \frac{dW_{d \to u}}{d\Omega} \right) \tag{23}$$

$$\frac{dW^{\text{down}}}{d\Omega} = \frac{1}{2} \left(\frac{dW_{d \to d}}{d\Omega} + \frac{dW_{u \to d}}{d\Omega} \right) \tag{24}$$

Any asymmetry in the two currents is best characterized by ensemble averaged asymmetry parameter $\langle A \rangle$ associated with the un-polarized target atoms, defined by

$$\langle A \rangle = \frac{\left(\frac{dW^{\text{up}}}{d\Omega} - \frac{dW^{\text{down}}}{d\Omega} \right)}{\left(\frac{dW^{\text{up}}}{d\Omega} + \frac{dW^{\text{down}}}{d\Omega} \right)} \tag{25}$$

In Figure 1 we show the result of calculations for the asymmetry parameter $\langle A \rangle$ as a function of the polar angle of the emitted electrons, at laser intensity, $I = 17.55e + 10^{16}$ W/cm^2 for $\omega = 3au$ (outer curve) and $\omega = 4au$ (inner curve). Remarkably, unlike the asymmetry parameter for the Fano effect [13], the two curve reveal a strong dependence of $\langle A \rangle$ on field frequencies at all angles. The absolute size of $\langle A \rangle$ is larger for the low frequencies at all angles. The "peak values" are seen on the plane of polarization ($\theta = 90^o$) and independent of intensities). $\langle A \rangle$ In Figure 1 for both frequencies $\langle A \rangle$ is mostly as large as 10^{-4} in magnitude but negative. Clearly, the negative sign indicates a dominance of the spin-down electron current over the spin-up current at all angles.

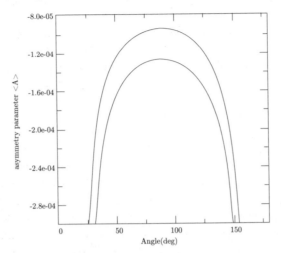

Figure 1. Frequency dependent ensemble averaged asymmetry parameter $\langle \mathbf{A} \rangle$ (see text for definition) vs. electron emission angle, $I = 17.55\mathrm{e} + 16\mathrm{W/cm}^2$, $w = 81.6\mathrm{eV}$ (outer curve) and 108.8eV (inner curve)

To understand the dominance we examine the corresponding spin-flip asymmetry parameter A. It is defined as the difference at a given angle between the spin-flip rates in the opposite directions scaled by the total rate at the same angle,

$$A = \left(\frac{dW_{u \to d}}{d\Omega} - \frac{dW_{d \to u}}{d\Omega} \right) \bigg/ \left(\frac{dW^{\mathrm{up}}}{d\Omega} + \frac{dW^{\mathrm{down}}}{d\Omega} \right) \tag{26}$$

At both the intensities A is positive implying clearly that the $u \to d$ flip rate is greater than the $d \to u$ flip rate (eqs. 18, 19). We note that the angular dependence of A and $\langle \mathbf{A} \rangle$ in Figures 1 and 2 are rather similar (but for the opposite sign). Also the magnitude of the asymmetry A is comparable to that of $\langle \mathbf{A} \rangle$. Thus the dominance of the $\mathbf{u} \to \mathbf{d}$ spin-flip rate itself over the $\mathbf{d} \to \mathbf{u}$ rate (for the present choice of the field polarization) leads to the negative values of $\langle \mathbf{A} \rangle$ despite the 50/50 weighting of the two initial spin states in the latter case. [8].

More so angular ionization rate $\mathbf{d} \to \mathbf{d}$ is larger than $u \to u$ at all angles. For $\mathbf{u} \to \mathbf{u}$ (or $\mathbf{d} \to \mathbf{d}$) the "peak values" are seen to occur at a somewhat smaller angle from the plane of polarization (Figure 3). The rates are seen to depend on frequency and intensity. This can be seen from the expression of the momentum transfer $\tilde{\mathbf{q}}$ which depends on $\tilde{\kappa}$, and on intensity via $(\mathbf{A_0/c^2}) = \mathbf{I}/\omega^2$. Peak value of of the difference between $\mathbf{d} \to \mathbf{d}$ (dd) curtrent and $\mathbf{u} \to \mathbf{u}$ (uu) current (i.e.dd-uu) for intensity $I = 17.55\mathrm{e}+16\mathrm{W/cm.sq}$ at frequencies 81.6ev (3au) and 108.8ev (4au) are respectively 1.1842e+10 and 4.9579e+9 per sec (Figure 4). The rate for spin-flip ionization current $\mathbf{u} \to \mathbf{d}$ (ud) is maximum in the forward direction (Figure 5). While du is identically zero (19).

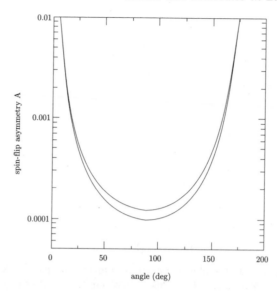

Figure 2. Spin-flip Asymmetry A (see text) vs. angle of electron emission at $I = 17.55$ e+16W/cm^2, w= 81.6eV (outer curve) and 108.8eV (inner curve)

Perhaps the simplest way to examine the necessity or otherwise of the retardation effect for the spin-flip process is to put the light propagation vector $\tilde{\mathbf{k}}$ identically equal to zero in the transition matrix elements $\mathbf{M}_{u \rightarrow d}$ in eq. (18). The spin-flip amplitude does not vanish in the limit of zero retardation or if the magnetic component of the incident laser field is neglected. It is to note that in the present case there is no spin -orbit interaction in the initial state (ground s-state) or in the final state as approximated by the plane wave Dirac-Volkov continuum state. So one may inquire about the mechanism for the finite spin -flip transition probability in intense fields in the absence of spin-orbit interaction. We note that the (eqn.18) for spin-flip amplitude depends on $\mathbf{m_2}$ and $\mathbf{g(q)}$ both of which arise from the "weak" components of the Dirac spinor of the free electron and the ground state of the Dirac H-atom, respectively. Hence they certainly go beyond the usual Pauli mechanism of coupling of the external magnetic field to the spin (magnetic moment) of the electron The equation for the $\mathbf{u} \rightarrow \mathbf{d}$ current shows that non vanishing coupling occurs through the vector product of the polarisation and the electron momentum $\tilde{\mathbf{p}}$. This along with the outer factor $\frac{\mathbf{A_0}}{\mathbf{c}} \simeq \frac{\mathbf{E_0}}{\mathbf{w}}$; where $\mathbf{E_0}$ is the electric field amplitude of the laser field in the laboratory, shows that the coupling depends on factors of the form $\mathbf{E_0} \times \tilde{\mathbf{p}}/\mathbf{c} \approx \tilde{\mathbf{B}}'$ (a.u.) where $\vec{B}'$ is, in fact, an effective magnetic field seen by the electron in its own frame of reference. It arises from the Lorentz transformation (e.g., [9]) of the electric field E_0 of the laser in the laboratory, into an effective (or "motional") magnetic field in the rest frame of the electron moving with a momentum $\tilde{\mathbf{p}}$ in the laboratory. Therefore the dominant mechanism that leads to the spin-flip transition in intense laser fields is the coupling of the motional magnetic field $\tilde{\mathbf{B}}'$ with the spin $\tilde{\sigma}$ or magnetic moment $=1/4c^2$ (a.u.) of the electron. This is rather analogous but not identical to the Lorentz transformation of the electric field associated with the static atomic potential $V(r)$ into an effective magnetic field and the resulting spin-orbit interaction responsible for the well-known Fano effect [10] observed in the perturbative domain of intensity [11].

It is worth noting from the point of view of controlling the relative dominance of the up $\rightarrow$ down or down $\rightarrow$ up spin-flip probability (and hence of the spin-down or spin-up electron currents) that their magnitudes can be reversed from outside, for example, by changing the helicity of the incident light. This can be seen by replacing the right circular polarization vector $\vec{\varepsilon}(\xi = +\pi/2)$ by the left circular polarization vector $\vec{\varepsilon}'$ $(\xi = +\pi/2)$ in Eqs. (17)–(20) and observing that the following transformations of the amplitudes hold: $\mathbf{M_{u \rightarrow u}} \rightarrow \mathbf{M_{d \rightarrow d'}}, \mathbf{M_{d \rightarrow d}} \rightarrow \mathbf{M_{u \rightarrow u}}, \mathbf{M_{u \rightarrow d}} \rightarrow -\mathbf{M_{d \rightarrow u}}$ and $\mathbf{M_{d \rightarrow u}} \rightarrow -\mathbf{M_{u \rightarrow d}}$. Hence, the spin-flip rates would exchange their magnitudes and the asymmetries $\langle \mathbf{A} \rangle$ would change its sign on changing the helicity of the photons from the right circular to the left circular polarization.

Before concluding, we may recall that the magnitude of the asymmetry parameters $\langle \mathbf{A} \rangle$ and A, in the cases explicitly considered above, are of the orders of $\mathbf{10^{-4}}$ for the near-infrared wavelength. These values lie well above the threshold efficiency $\approx 24 \times 10^{-4}$ of currently available spin analyzers in the laboratory (e.g., [12, 13]). The asymmetry in spin currents can give an estimate of the amount of ortho and para hydrogen present in an ensemble of hydrogen atoms by experimentally measuring the spin polarized ionization currents.

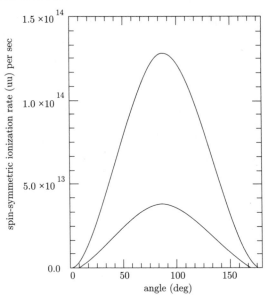

Figure 3. Spin-symmetric ionization rate $u \rightarrow u$ per sec vs. angle of electron emission at $I = 17.55\mathrm{e}{+}16\mathrm{W/cm}^2$, $w = 81.6\mathrm{eV}$ (outer curve) and $108.8\mathrm{eV}$ (inner curve)

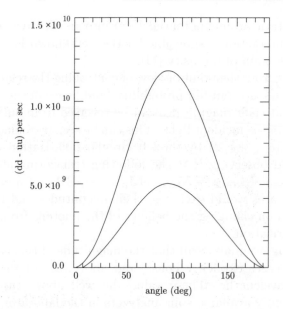

Figure 4. Difference of spin-symmetric ionization rate per sec between $d \to d$ and $u \to u$ vs. angle of electron emission at $I = 17.55\mathrm{e}{+}16\mathrm{W/cm}^2$, $w = 81.6\mathrm{eV}$ (outer curve) and 108.8eV (inner curve)

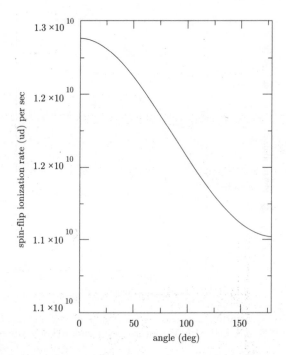

Figure 5. Spin-flip ionization rate (ud) $u \to d$ per sec vs. angle of electron emission at $I = 17.55\mathrm{e}{+}16\mathrm{W/cm}^2$, $w = 81.6\mathrm{eV}$

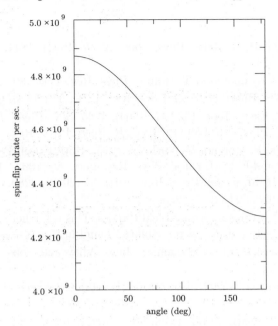

Figure 6. Same as Figure 5 for $w=108.8$eV

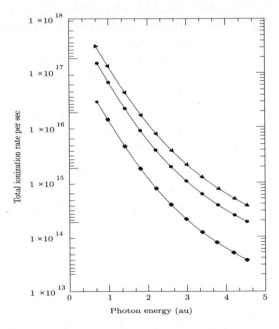

Figure 7. Total ionization rate (average over the initial spin and sum over final spins) per sec Vs laser photon energy (au) at intensities $I=10$au (triangle),5au (circle), 1au (diamond). 1au of energy=27.2eV, 1au of intensity=3.51e+16W/cm^2

REFERENCES

1. D. P. Crawford and H. R. Reiss, Phys. Rev. A, 50 (1994), 1844; H. Reiss, J. Opt. Soc. Am. B, 7 (1990), 574.

2. N. Milosevic, V. P. Krainov and T. Brabec, Phys. Rev. Lett., 89 (2002), 193001.

3. F.H.H. Faisal and S. Bhattacharyya, Phys. Rev. Lett., 93 (2004), 053002.

4. L.V. Keldysh, Zh. Eksp. Teor. Fiz., 47 (1964), 1945 [Sov. Phys. JETP, 20 (1965), 1307]; F. H. M. Faisal, J. Phys. B6, L89 (1973); H. R. Reiss, Phys. Rev. A, 22 (1980), 1786.

5. A. Becker, L. Plaja, P. Moreno, M. Nurhuda and F. H.M. Faisal, Phys. Rev. A, 64 (2001), 023408; A. Becker and F. H.M. Faisal, Phys. Rev. Lett., 89 (2002), 193003; J. Muth-Bohm, A. Becker and F. H. M. Faisal, Phys. Rev. Lett., 85 (2000), 2280.

6. D.M.Wolkow, Z. Phys., 94 (1935), 250.

7. F. H.M. Faisal, in Lectures in Strong-Field Physics, This is further supported by the almost equal values of $u \to u$ and $d \to d$ probabilities found in these cases.

8. J. S. Jackson, Classical Electrodynamics, John Wiley and Sons, New York, 1975, 2nd edition.

9. U. Fano, Phys. Rev., 178 (1969), 131; see also P. Lambropoulos and M. R. Teague, J. Phys. B, 9 (1976), 587.

10. For weak laser fields contribution of spin-orbit interaction of the order of $-\frac{1}{r}\frac{dV(r)}{dr}\vec{r} \times \vec{p}/2c$ (a.u.) has to be added to $\vec{B}'$ for quantitative accuracy. For strong laser fields it is negligible compared to $\vec{B}'$ and hence cannot affect the numerical results qualitatively.

11. M. Getzlaff *et al.*, Rev. Sci. Instrum., 69 (1998), 3913.

12. J. Kessler, Polarized Electrons, Springer-Verlag, Berlin, 1985, 2nd edition.

Atoms and Molecules in Laser and External Fields

Editor: Man Mohan

Two-Photon X-Ray Transitions in Simple High-Z Atomic Systems

P. H. Mokler

GSI, Planck-Str. 1, D-64291 Darmstadt, Germany and University of Giessen, Germany

INTRODUCTION

The structure of atoms is typically probed in experiments by radiative transitions giving direct access to binding energy differences of the involved states. In theory atomic binding energies are deduced from matrix elements where a wave function is weighted by the Hamiltonian, i.e. it is multiplied from both sides by the same wave function. Here, variational methods minimizing the binding energy of a level are often applied and provide accurate results. These methods are mainly sensitive to the radial structure of the wave functions. In recent years precise determination of binding energies revealed also for the heaviest few-electron systems detailed insight into the atomic structure for the case of strong central fields, determining there even higher order quantum electrodynamical effects, cf. Ref. [1]. In contrast to this research line, lifetime studies are more responsive to the exact shape of the density distributions of the electrons - in particular also to the angular part of the wave functions. Here, the overlap of the two wave functions involved weighted by the operator for the interaction determines the transition rates. This is equivalently true for both line shapes and for line widths. For heavy atomic systems the inner shell transitions are extremely fast and thus, for these fast allowed transitions, lifetimes or line widths are extremely hard to measure, if not impossible. Hence, in heavy atomic systems lifetimes or line widths are only accessible to experiments by metastable states or by "forbidden transitions". In those cases, a direct access to the detailed structure of the wave functions involved is possible even in very heavy atomic systems.

For instance, in He-like systems a direct transition from the "metastable" $1s2s\,^1S_0$ state to the $1s^2\,^1S_0$ ground state is totally forbidden by the selection rules. This state can only decay by the emission of two photons to the ground state, i.e. by a 2E1 transition. To some extent, the same is true for the excited $2s\,^2S_{1/2}$ state in H-like ions; however, there additionally a direct $M1$ decay to the $1s\,^2S_{1/2}$ ground state competes with the

2E1 transition. For heavy systems the decay is still fast in both the cases. However, the shape of the two-photon spectra is experimentally accessible and gives information on the structure of the heavy atomic system. For illustration, the term diagram for a heavy He-like ion together with the ground state transitions (left side) as well as their rate dependences (right side) are shown in Figure 1. (An overview of the rate dependences is given e.g., in Ref. [2] and in the literature cited there.)

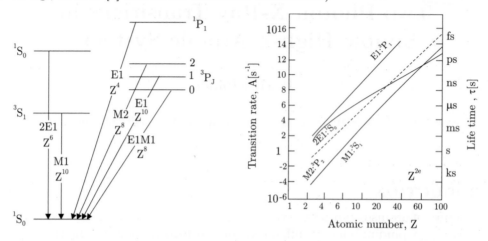

Figure 1. Level diagram for a He-like ion (left side); the general Z dependences for the ground state transitions are indicated. In the right hand diagram the transition rates and lifetimes for possible ground state transitions are given, cf. [2]

The two-photon decay was treated theoretically first by Göppert-Mayer [3], more than 75 years ago: The 2E1 decay rate is determined by a summation over all two-photon transitions possible via all the intermediate P states of the atomic system, where the sum energy of the two photons corresponds to the total binding energy difference between the initial and final state. The summation includes all relevant bound and continuum states. The differential transition probability $W_{2\gamma}$ has the form [3]:

$$dW_{2\gamma}/d\omega_1 = \{\omega_1 \cdot \omega_2/(2\pi \cdot c)^2\} \cdot |M_{2\gamma}|^2 \cdot d\Omega_1 \cdot d\Omega_2 \tag{1}$$

where ω_j is the energy and $d\Omega_j$ the solid angle for the jth photon and the transition energy ω_0 satisfies the energy conservation $\omega_0 = \omega_1 + \omega_2$. The second order matrix element $M_{2\gamma}$ is given by the expression:

$$M_{2\gamma} = \varepsilon_1 \cdot \varepsilon_2 \Sigma\{\langle 1^1S_0\|R_{E1}(\omega_2)\|n\rangle\langle n\|R_{E1}(\omega_1)\|2^1S_0\rangle/(E_n - E_{2'So} + \omega_1)$$
$$+ \langle 1^1S_0\|R_{E1}(\omega_1)\|n\rangle\langle n\|R_{E1}(\omega_2)\|2^1S_0\rangle/(E_n - E_{2'So} + \omega_2)\}$$

Here, ε_j is the polarization vector for the jth photon and $R_{E1}(\omega_j)$ is the electric dipole operator. It has to be mentioned that the summation runs over all intermediate states n 0independent of whether they are empty or occupied [4].

Integrating finally over all possible photon energies yields the total transition rates that are sensitive to the structure of the entire atomic system. Measurements of lifetimes for the $1s2s^1S_0$ and the $2s^2S_{1/2}$ levels in He- and H-like ions, respectively, and their

Z dependences give therefore already important information on the total structure of heavy few-electron ions; for an overview see e.g. Refs. [5, 6]. The 2E1 transition rates increase strongly with Z - in first approximation with the sixth power of the atomic number, Z^6 - due to the increase in total transition energy, cf. Figure 1.

THE TWO-PHOTON CONTINUUM

In contrast to lifetime measurements, a determination of the whole two-photon decay spectrum will provide much more detailed information on the entire structure of He-like atomic systems. In this case, the shape of the measured two-photon continuum has to be compared with the energy differential transition probability, $dW_{2\gamma}/d\omega_1$, described above. As the two photons are indistinguishable, the spectrum is symmetric in the photons, i.e. it is mirror symmetric around the mid point at half the total transition energy, $\omega_0/2$. For convenience in comparing the spectral shape from different atomic systems, the total transition energy is normalized to "1" with the fractional energies of the two photons

$$f_j = \omega_j/\omega_0 \text{ and } f_1 + f_2 = 1.$$

In this representation the differential transition probability for the two-photon decay can be approximated by a product of a simple "phase space factor", $f_1 \cdot (1 - f_1)$, and a "structure factor", c.f. Ref. [7]:

$$dW_{2\gamma}/d\omega_1 \propto f_1 \cdot (1 - f_1) \bullet |M_{2\gamma}|^2 \tag{2}$$

The phase space factor has just the shape of a parabola and the structure factor is solely determined by the square of the energy differential matrix element $M_{2\gamma}$. The structure factor contains all the information on the entire atomic system and varies considerably with atomic species [8-10]. A survey on the two-photon decay in heavy ions and atoms was given recently in Refs. [11-13]. There special emphasis was put to the change of the spectral shape with atomic number Z. Moreover, the two-photon decay in singly K-shell ionized heavy atomic systems is also included. These quasi He-like atomic systems were investigated in detail by Ilakovac and his group [14-17]. In the following, we concentrate only on heavy true He-like ions.

In Figure 2 the shape of the two-photon distribution and its change with Z is presented for He-like ions. The structure factors for He, Ni^{26+} and Au^{77+} are given along with the form of the parabola according to calculations by Derevianko and Johnson [10]. The structure factors for He and Au^{77+} are pretty similar whereas that for Ni^{26+} deviates considerably broadening the parabolic shape of the spectrum.

Considering just the widths (FWHM) of the continuous two-photon spectra alone, the structure factor broadens in general the distributions in a way that depends on the system. In Figure 3 the FWHM values (in relative units of ω_0) of the two-photon distributions are schematically depicted as function of Z. For H-like systems the FWHM of the continuum ("fractional" width) starts at a value of 0.84 for H and decreases with increasing atomic number due to relativistic effects.

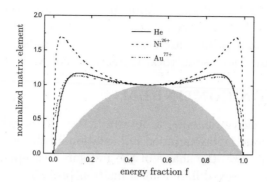

Figure 2. The shape factors for the 2E1 decay continuum in He-like ions. The structural factor for He, Ni^{26+} and Au^{77+} are given as normalized matrix elements (squared) along with the parabolic shape of the phase-space factor (rim of the shaded area) as a function of the fractional photon energy, acc. to [10]

(Excluding relativistic effects would yield a straight line, see the dashed line in the figure). For $H-$ like U the FWHM is about 0.71, which is also the value for a pure parabola, cf. Ref. [9]. For He the FWHM starts at 0.75, a narrower width compared to that of H. This narrowing is caused by the electron-electron interaction. Due to the relative decrease of this interaction with atomic number Z compared to that of the central potential the width for He-like ions widens first towards that of H-like systems. Around the mid-Z region the relativistic effects turn the tendency back, cf. Ref. [10]. Finally the width for He-like U approaches that of H-like U - an indication that $H-$ and He-like atomic structures approach each other for the heaviest atomic species. However, these widths are only an integral measure of the structure, more detailed information can be gained by studying the true shape of the two-photon continuum. On this issue we will concentrate for heavy He-like ions below.

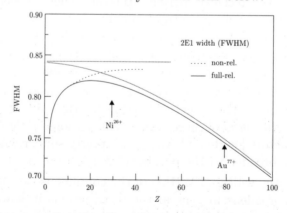

Figure 3. Fractional FWHM values (i.e., in relative units of ω_0) of the two-photon continuum are sketched as function of the atomic number Z. The upper two curves are for H-like systems and the lower ones for He-like systems; the dashed curves give pure non-relativistic approaches and the full curves include relativistic effects, respectively

EXPERIMENTAL TECHNIQUES

In order to produce He-like heavy ions the acceleration-stripping technique is normally used, where high velocities are needed for heavy ions. To efficiently produce He-like species of, for instance, Ni and Au typical energies around 10 and 100 MeV/u are needed (compare the binding energies for the K electrons of about 9 and 85 keV), respectively. Then, the He-like fast ions are magnetically selected before they are excited in a second "exciter foil" with a certain probability to the $1s2s^1S_0$ state. The lifetime of this state is in Ni^{26+} 154 ps [18] and in Au^{77+} only 0.32 ps [19]. The radiative decay of these excited states is detected behind the foil by at least two solid-state x-ray detectors (Si(Li) and Ge(i) detectors, respectively) looking face-to-face perpendicular to the downstream ion beam. The two photons of the decay are preferentially emitted 1800 apart and are detected in coincidence.

Due to the short lifetimes of the Au ions these ions decay close to the exciter foil despite their high velocities. Moreover, in this relativistic velocity regime the Lorentz transformation of the photon emission angle from the fast ion system into the laboratory frame leads to a forward tilt of the "900^0" observation angle [20, 21]. The experimental arrangement for the Au^{77+} case is given on the left side in Figure 4. In order to be able to compensate for the Doppler effect of the emitted photons at different observation angles one of the Ge(i) detectors is granularly divided into stripes. For a precise determination of the true two-photon spectrum, emission angles, detector efficiencies, photon transmission losses and electronic coincidence efficiencies have to be determined with high accuracy. Figure 4 gives on the right side for the Au experiment a event plot of coincident photons measured with both the detectors. The ridge along the diagonal line describes the wanted two-photon continuum convoluted with all the relevant efficiencies.

RESULTS

Due to the complicated efficiency response of the detection system, the measured coincident two-photon spectra cannot be compared directly to theory. Instead of unfolding the measured spectra with all the uncertainties involved, the theoretical spectral distributions are normally convoluted with all the relevant efficiencies and then compared directly with the experimental ones. This convolution is done best by Monte Carlo simulations starting from the theoretical spectral distribution and then including all the experimental correction factors, cf. Ref. [6]. This method was successfully introduced for the case of the two-photon continuum in He-like Kr^{34+} ions [6]. There a clear change in the relative spectral distribution for the two-photon decay compared to the one for atomic He was established. Unfortunately, the systematic errors are still somewhat large. A further improvement in accuracy was achieved for Ni ions by comparing the two-photon decay for H− and He-like species, Ni^{27+} and Ni^{26+}, respectively [7]. Assuming that the spectral distribution for H− like ions can be treated correctly all the experimental efficiencies can be extracted with high accuracy; and then, the spectral distribution for the He-like case can be compared to theory yielding the most precise determination for the two-photon continuum in heavy He-like ions done up to now. However, at this Z range the relativistic effects on the spectrum are still too small to be uniquely detected. Relativistic [10] and non-relativistic [22] calculations still deliver here quite similar distributions.

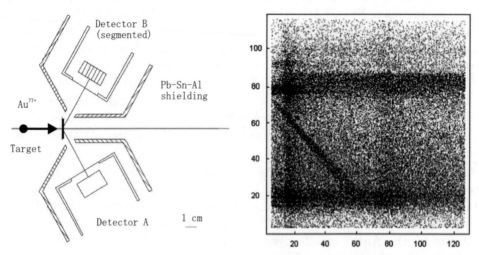

Figure 4. The experimental arrangement for the Au experiment (106 MeV/u Au^{77+} ions excited in an Al foil) is given in the left part. The right part depicts a event plot for the coincident emission of two photons (detector A vs. detector B, photon energies in keV). Along the diagonal ridge in the event plot the sum energy of both photons is constant and corresponds to the total transition energy of the 2E1 decay [20, 21]

For the heaviest investigated system, He-like Au^{77+} ions, relativistic effects are strong. Hence, a comparison of the experimental spectrum with the simulation is displayed at the top of Figure 5 [21]. Within the error band for the simulation and the experimental error bars, good agreement results between measurement and fully relativistic theory. For comparison, a corresponding simulation for the non-relativistic case of Ar^{16+} ions is also given (dashed line). At the bottom the residues of the experimental data with respect to the results of the simulations are given. It is evident that the non-relativistic approach is at variance with the experimental findings, only the fully relativistic approach is in agreement. (The slight right-left asymmetry in the data may indicate to a small angular misalignment in the experiment.)

Taking out the trivial phase space factor from the continuous spectrum one obtains the energy dependent square of the matrix element, see Figure 6 (bottom part). For comparison the normalized square of the theoretical matrix elements are shown for He-like Ni and Au ions along with the experimental data for Au^{77+}. The difference in the distributions demonstrates quite clearly the relativistic influence, whereby all intermediate ^3P states are already included in the calculations [10]. The experimental points evidently favor the fully relativistic Au distribution; a χ^2 test gives values of 3.48 and 7.35 for the Au and Ni predictions, respectively. However, it is also evident, that additional and more precise measurements are needed for a more refined comparison with theory; those experiments are presently in progress.

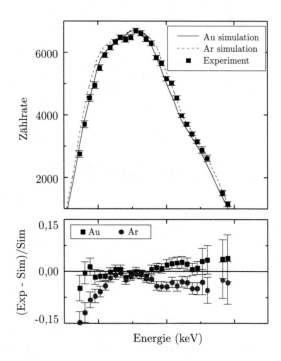

Figure 5. The x-ray continuum for the two-photon emission from He-like Au^{77+} ions (top part) [21]. The measured data points are shown along with Monte Carlo simulations one for fully relativistic Au^{77+} ions (full line) and one for non-relativistic Ar^{16+} ions (dashed line), according to Ref. [10]. At the bottom the residues of the experimental data points compared to the simulations are given

For comparison the results for the "non-relativistic" case of Ni ions are shown to scale at the top insert in Figure 6 [7]. At that (low-) medium atomic number ($Z = 28$) the experiment with its really small uncertainties cannot distinguish between non-relativistic and full relativistic calculation as both gives almost the same result. However, a pronounced variance to both the expectations for the low Z case of He ($Z = 2$) and the strong relativistic case of Au ions ($Z = 79$) is evident.

SUMMARY AND OUTLOOK

Both experimentally and theoretically, important steps towards a better understanding of the entire atomic structure of He-like heavy ions were made recently by investigating the two-photon decay continuum associated with the excited $1s2s\,^1S_0$ state. Nowadays, non-relativistic approaches [22] valid up to the medium Z region (around 30) are complemented also for heavy systems by fully relativistic calculations [10] including all intermediate ^3P states. Measurements in the medium Z region still compare quite well with both calculations as the relativistic effects are not yet dominant there [6, 7].

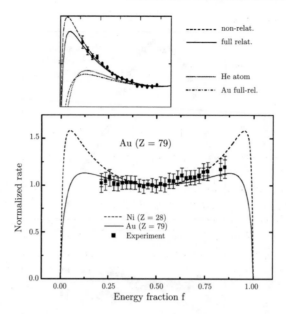

Figure 6. The normalized square of the energy dependent matrix element (rate) extracted from the experiment [21] is compared to the fully relativistic theory for Au^{77+} and Ni^{26+} [10] (full and dashed lines in the bottom part, respectively). In the insert at the top the corresponding results are shown for the non-relativistic case of Ni ions [7] along with different predictions [10, 22]

For a true relativistic system, Au^{77+}, the first experiment clearly favors the fully relativistic calculations. Higher accuracies in future experiments using the described beam foil excitation technique are anticipated for the near future.

Moreover, also alternative methods may be utilized to produce the meta-stable He-like $1s2s^1S_0$ state at moderate relativistic ion velocities. In a dedicated experiment it was found that selectively ionizing the K shell of Li-like ions leads with high efficiency to the $1s2s^1S_0$ excited state, cf. Figure 7 [23]. There Li-like U^{89+} ions colliding with N gas were ionized in the K-shell leaving the L shell undisturbed; the x-ray emission was measured in coincidence with the singly up-charged U^{90+} ions. In the particle / x-ray coincidence spectra displayed in Figure 7 only the $M1$ and $2E1$ decay modes are observed for that case. This new experimental method may give improved access to this area.

Additionally, with the last generation synchrotron radiation facilities new methods are at hand to produce singly K ionized atoms of any atomic number [24]. These singly K ionized atoms decay to a small fraction of the time via the two-photon decay branch and thus give information on the entire atomic structure of multi-electron systems which finally can be compared to the corresponding true He-like cases. First results from a synchrotron radiation experiment performed at APS in Argonne are already published for Au [24]. In the past, most of the data on inner shell two-photon decay in singly ionized systems were obtained from nuclear unstable species decaying by nuclear K

electron capture. This method was demonstrated and successfully applied by Ilakovac and his group [14-17]; see also the recent work in Ref. [25].

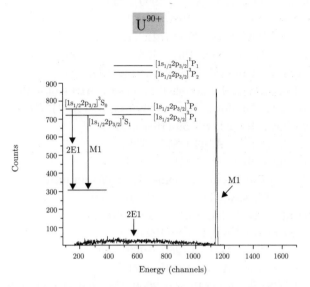

Figure 7. X-ray spectrum measured in coincidence with K-shell ionized U^{90+} ions from $U^{89+} \to N_2$ collisions at 217 MeV/u [24]. The relevant decay scheme is indicated at the top

The present results give a first insight into the two-photon decay of true relativistic — i.e., very heavy — He-like atomic systems. The shape of the two-photon spectra reflects the strucure of the entire atomic system including all occupied and non-occupied states. At present the two-photon decay seems to be the most promising way to get detailed access to the entire structure of heavy atomic systems. Further measurements, both at heavy ion accelerators and at modern synchrotron radiation facilities, will give unprecedented insight to the entire structure of He-like or quasi-He-like atomic systems in the strong field domain of high Z species.

Acknowledgements

The long-standing and excellent collaboration with the atomic physics group at Argonne, especially with R.W. Dunford, E.P. Kanter and colleagues - is highly appreciated. Moreover, without the close collaboration with colleagues from Krakow, Poland (A. Warczak) and within the GSI atomic physics group the progress in this field would not have been possible.

REFERENCES

1. P. H. Mokler, Radiation Physics and Chemistry, (2006), in print.
2. P. H. Mokler et al., Phys. Scripta T., 51 (1994), 28.
3. M. Göppert-Mayer, Ann. Phys. (Leipzig), 9 (1931), 273.
4. Y. B. Bannet and Freund, I., Phys. Rev. A, 30 (1984), 299.

5. R. Marrus and P. J. Mohr, Adv. At. Mol. Phys., 14 (1978), 181.

6. R. Ali et al., Phys. Rev. A, 55 (1997), 994.

7. H. W. Schäffer et al., Phys. Rev. A, 59 (1999), 245.

8. G. W. F. Drake, Phys. Rev. A, 34 (1986), 1182.

9. W. R. Johnson, Phys. Rev. Lett., 29 (1972), 1123.

10. A. Derevianko and W. R. Johnson, Phys. Rev. A, 56 (1997), 1288.

11. P. H. Mokler and R. W. Dunford, Fizika A, 10 (2001), 105.

12. P. H. Mokler et al., X-Ray and inner shell processes, AIP Conf. Proc., 652 (2003), 237.

13. P. H. Mokler and R. W. Dunford, Physica Scripta, 69 (2004), C1-9.

14. K. Ilakovac et al., Phys. Rev. Lett., 56 (1986), 2469.

15. K. Ilakovac et al., Phys. Rev. A, 44 (1991), 7392.

16. K. Ilakovac et al., Phys. Rev. A, 46 (1992), 132.

17. K. Ilakovac, Radiation Physics and Chemistry, (2006), in print.

18. R. W. Dunford et al., Phys.Rev.Lett., 62 (1989), 2809.

19. W. R. Johnson et al., Adv.At.Mol.Phys., 35 (1995), 255.

20. H. W. Schäffer, GSI-report Diss., 99-16 (1999).

21. H. W. Schäffer, et al., Phys. Lett. A, 260 (1999) 489.

22. G. W. F. Drake, Phys. Rev. A, 34 (1986), 2871.

23. Th. Stöhlker et al., GSI Scientific Report 2000, GSI-2001-1 (2001), 95.

24. R. W. Dunford et al., Phys. Rev. A, 67 (2003) 054501.

25. P. H. Mokler et al., Phys. Rev. A, 70 (2004) 032504.

Atoms and Molecules in Laser and External Fields
Editor: Man Mohan

Laser Spectroscopy and Epr Analysis of the Strong Ga Atoms/sio$_2$ Surface Interaction

S. Barsanti[1], M. Cannas[2], E. Favilla[1] and P. Bicchi[1]

[1] Department of Physics, University of Siena, Via Roma, 56-53100 Siena, Italy
[2] Department of Physical and Astronomical Sciences, University of Palermo, Via Archirafi, 36-90123 Palermo, Italy

1. INTRODUCTION

There is a long lasting tradition in the study of the products of the collisions between resonantly excited atoms in dense vapours of the semiconductor elements [1]. The attention was recently concentrated on Ga and ionization as well as autoionization proceeding from laser assisted collisions were reported [2,3]. One of the essential parameters to determine the cross section of these processes is the exact knowledge of the element atomic density at the temperature of the experiment. While performing these measurements, the density of the Ga atoms, when the vapour was confined in Herasil 3 quartz cells at temperatures $T \geq 900°$C, came out to be sensibly lower than the value expected from the corresponding saturated vapour pressure [4]. At the same time the fluorescence spectrum induced by the excitation with laser photons resonant with the fundamental Ga transition at 403.4 nm displayed some features that were not ascribable to any Ga fluorescence and, highly surprisingly, once manifested, were permanent. They would show up even if the cell was cooled down to room temperature and then was illuminated with the same photons. This behavior was interpreted as due to the strong interaction Ga/SiO$_2$ surface that produces a migration of the Ga atoms inside the silica matrix where they remain trapped. This effect changes the fundamental properties of the vapour such as the radiation trapping [4], but it also changes the silica optical characteristics which raises a considerable interest as this Ga-doped silica may have potential applications in the field of silica based opto-electronic devices where the optical properties of defects in silica are largely used [5]. For this reason we started a systematic study of different silica samples, both natural and synthetic, thermally treated in a Ga atmosphere and in an inert He atmosphere, using *Optical Absorption*

(OA) and *Laser Induced Fluoresce* (LIF) technique as well as the Electron Paramagnetic Resonance (EPR) one. The very first observations have been published [6], while in the following we report the updated results of these checks.

2. EXPERIMENTAL

We investigated 3 silica samples: a piece of Herasil 3 silica taken from the body of the cells used for 3 the laser assisted collisions experiment and 2 commercial silica samples ($5 \times 5 \times 1$ mm^3 sized): (a) natural wet Herasil 3 prepared by fusion of quartz in water-vapour atmosphere; it contains nearly 200 part per millions (ppm) by weight of OH groups and metallic impurities such as Al, Ge or alkali, totally of the order of 10 ppm as the unfused raw material and (b) synthetic wet Suprasil 1 made by vapour-phase hydrolysis of pure silicon compounds such as $SiCl_4$; it contains 1000 ppm of OH but it is virtually free from metallic impurities. Samples of each silica specimen were submitted to two different thermal treatments in which they are heated at $1050°$C for 8 hours either in a Ga atmosphere or in a He one in the apparatus drawn in section A of Figure 1. The oven is a stainless steel heat pipe oven (HPO) [7] whose central part is heated by a current flowing in electric wires and whose temperature is kept fixed by a termocouple + thermostat system within "$1°$C. The samples are placed at the center of this hot part where the Ga or He atmosphere can be created.

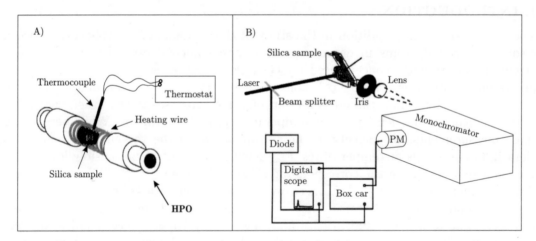

Figure 1. Heating (A), and LIF (B) apparatus

The essential of the LIF apparatus is sketched in section B of Figure 1. The thermally treated silica samples are illuminated by pulsed laser photons of different wavelengths in the UV, produced by a dye-laser pumped by the third harmonic of a Nd^+-Yag laser. The induced fluorescence is collected via -1 an iris/lens system, dispersed by a Jobin-Yvon 1000 monochromator with 20 cm resolution and detected by a photomultiplier whose output feeds a box-car integrator for frequency resolved spectra or a digital scope for time resolved ones. A small portion of the laser beam triggers the whole experiment. The OA spectra in the visible-UV range (600-200 nm), were taken with a double beam spectrometer whose acquisition parameter were set to: bandwidth 2nm, time response 1s and scan speed 1.6 nm/s.

EPR spectra were carried out by a spectrometer (Bruker EMX) working at $n_o = 9.8$ Ghz and acquired by scanning the magnetic field over the range 150-550 mT with a microwave power of 200 mW and a modulation field with peak-to-peak amplitude 0.4 mT at 100 kHz. The LIF, OA and EPR spectra are all taken at room temperature. As the Herasil 3 sample taken from the cells shows exactly the same behaviour of the commercial Herasil 3 sample, in the following it will not be distinguished anymore.

3. RESULTS AND DISCUSSION

3.1. LIF spectroscopy

The LIF spectra detected in both Herasil 3 and Suprasil 1 samples after thermal treatment in Ga atmosphere are shown in Figure 2(a, b, c) for the three different excitation wavelengths 406.4 nm, 403.4 nm and 389.7 nm respectively. The Ga inclusions inside the silica matrix are testified by the permanent presence of peaks 1-4 in the spectra, but their behaviour testifies also of the different origin of some of them. In fact peaks 1, 2 and 3 show a typical Raman scattering aspect as they show up after the first excitation, Figure 2(a), and when the laser is detuned they shift of the same quantity as evident in Figure 2(b) and 2(c). Further support to their Raman scattering origin comes from their intensity that shows a linear dependence on the laser power density and also from their time resolved analysis that indicates they have the same temporal behaviour of the exciting laser photons. Peaks 4 manifests a totally different dependence on the excitation wavelength. It shows up in Figure 2(a) and if the laser is slightly detuned it remains present without any shift, Figure 2(b). If the laser is detuned considerably, it disappears, Figure 2(c).

We believe the Ga migration inside the silica produces both structural modifications in the SiO_2 matrix which are responsible for the Raman scattering spectra, and color centers which emit their typical fluorescence only if excited within their absorption band. The literature reports of both emission bands due to the interaction of Al (an element of the same group of Ga) atoms with fused silica [8] as well as Raman spectra of matrix-isolated Ga clusters [9] endorse our hypothesis as well as x some preliminary observations of a similar behaviour at In/SiO_2 boundary surface [10].

The fluorescence features that appear evident in all the spectra are totally absent both in brand new Herasil 3 and Suprasil 1 samples and in Herasil 3 and Suprasil 1 samples treated in He atmosphere, if illuminated by the same radiations.

3.2. OA and EPR Analysis

The peculiar role played by Ga in inducing structural modifications and/or color centers inside the silica matrix is also confirmed by looking at the EPR and OA spectra of the samples after their thermal treatment.

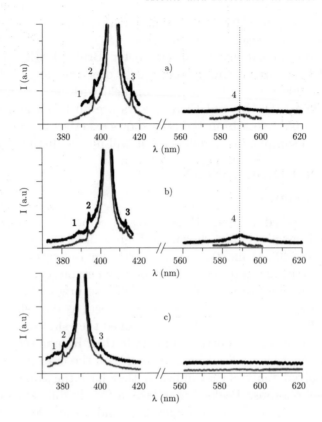

Figure 2. LIF spectra of Herasil 3 (black) and Suprasil 1 (gray) taken at room temperature after thermal treatment in Ga atmosphere following excitation at 406.4 nm (a), 403.4 nm (b) and 389.7 nm (c)

Figure 3 plots the comparison between the EPR spectra obtained in Herasil 3 and Suprasil 1 samples heated in Ga (trace 2) or He (trace 1) atmosphere. It is evident in both silica specimens that the thermal treatment with Ga induces a broad EPR lineshape extending over a few hundred mT, that is centered around the spectroscopic splitting factor $g = h_0/B2$, close to the factor of the free electron, h and 0 being the Planck constant and the Bohr magneton, respectively. A first value of the concentration of the active paramagnetic centers is derived to be about 10^{18} spins/cm^3, by comparing the double-integrated ESR spectra with that of Si-E centers, in a reference sample where their absolute density was measured with an accuracy of $\pm 20\%$ by spin-echo technique [11].

The OA spectra detected in Suprasil 1 samples heated in Ga and in He atmosphere, respectively, are shown in Figure 4 where it is evident the growth of a composite OA profile extending down to 450 nm in the Ga-heated sample. Similar effects, peculiar to Ga, were also observed in the Herasil 3 samples. These results allow to evidence that the wide broadening of both OA and EPR curves is consistent with the amorphous nature of silica matrix and then with the site-to-site non equivalence of defects randomly oriented with respect to the environment. This means that the defects interact with the matrix and explore their inhomogeneous properties. However, the origin of the OA and EPR signal and their relationship is actually an open question.

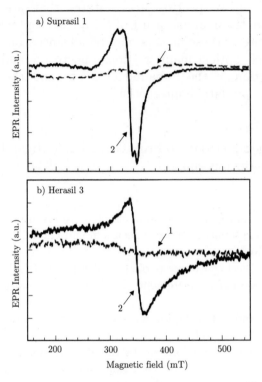

Figure 3. EPR spectra detected in Suprasil 1and Herasil 3 silica samples heated for 8 hours at a temperature of 1050° C in He (1) or Ga (2) atmosphere

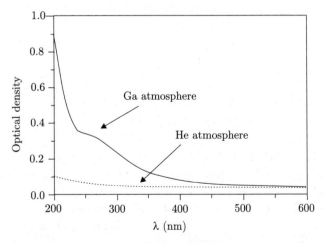

Figure 4. Comparison of OA spectra detected in samples of synthetic silica Suprasil 1 kept for 8 hours at a temperature of 1050° C in Ga or He atmosphere

4. CONCLUSIONS

We have reported the experimental evidence of thermally activated Ga atoms migration inside the silica structure which induces the formation of optically and EPR active defects responsible for new spectroscopic features that are of great interest for potential applications in the manufacturing of silica based optoelectronic devices. Some hypothesis on the nature of these defects may be advanced but work is in progress to better characterize their optical manifestations and to try and understand the exact extent of the localization of the unpaired electrons and their relaxation processes involving spinspin and spin-lattice interactions.

Acknowledgments

The authors express their gratitude to Prof. Geraldo Alzetta for the several enlightening discussions and the continuous support.

REFERENCES

1. P. Bicchi, S. Barsanti and E. Favilla, Products of collisions in presence of resonant photons in dense vapours of the semiconductor elements in and Ga in, Recent Res. Devel. Physics, 2004, 5, 839, TRN, Kerala.
2. S. Barsanti and P. Bicchi, Energy-pooling ionization (EPI) in Ga vapour: electronic detection and cross section measurement, J. Phys. B: At. Mol. Opt. Phys., 2001, 34, 5031.
3. P. Bicchi, S. Barsanti and E. Favilla, Laser assisted collisional population of Ga I autoionizing levels, Las. Phys., 2004, 14, 144.
4. P. Bicchi and S. Barsanti, Quenching of the radiation trapping in a dense Ga vapour in a quartz cell induced by atom/surface interaction, Rad. Phys. Chem, 2003, 68, 91.
5. R.A.B. Devine, J. P. Duraud and E. Dooryhée, Structure and imperfections in amorphous and cristalline dioxide, 2000, J. Wiley & Sons, New York.
6. S. Barsanti, M. Cannas, E. Favilla and P. Bicchi, Spectroscopy of an optical excited SiO surface, Rad. Phys. Chem, 2, 2006, in press.
7. I. C. Finlay and D. B. Green, Heat pipes and their instrument applications, J. Phys. E., 1976, 51, 1026.
8. J. H Schulman and W. D.Compton, Color Center in Solids, 1962, Pergamon, London.
9. F. W. Froben, W. Schulze and U.Kloss, Raman spectra of matrix-isolated group IIIA dimers: Ga_2, In_2, Tl_2, Chem. Phys. Lett., 1983, 99, 500.
10. P. Bicchi, C. Marinelli, E. Mariotti, M. Meucci and L. Moi, Radiation trapping and vapor density of indium confined in quartz cells, Opt. Commun., 1994, 106, 197.
11. S. Agnello, R. Boscaino, M. Cannas and F. M. Gelardi, Instantaneous diffusion effect on spin-echo decay: experimental investigation by spectral selective excitation, Phys. Rev. B, 2001, 64, 174423.

Atoms and Molecules in Laser and External Fields

Editor: Man Mohan

Copyright © 2008, Narosa Publishing House, New Delhi, India

1s-2s Raman Transitions in Hydrogen-Like Atoms

Sara Fortuna[1] and Naseem Rahman[1,2]

[1] Department of Chimical Sciences, University of Trieste, 34100 Trieste, Italy
[2] European Institute for Advanced Studies, EIAS Hamburg-Paris-Trieste, Via Giorgieri, 1-34100 Trieste, Italy

1. INTRODUCTION

Recent development of new experimental techniques allow us to study highly ionized atoms, e.g. high hydrogenic atoms.

It is possible, for example, to obtain H-like uranium ions[3] and, since 1977[4], it has been possible to trap a single ion in a Penning trap and study its characteristics: in 2005 a single $40Ca^{19+}$ was studied[5]. It is thus possible to study atomic properties as a function of their atomic masses (Z), considering constant all the other parameters.

In this paper we discuss the Raman-like effect for H-like atoms with masses from 1 to 92. This process is a 2 photon process where one photon is adsorbed and another photon is emitted from the system.

2. CALCULATIONS

For the Hydrogen-Like atoms, exact non-relativistic calculations can be done.

In the framework of the dipole approximation, the interaction of radiation with the electron in the r-gauge takes the following form in mks units:

$$V = -eE_1\hat{e}_1 \cdot \vec{r}\cos\omega_1 t - eE_2\hat{e}_2 \cdot \vec{r}\cos\omega_2 t, \qquad (2.1)$$

where E_1 and E_2 are the electric fields of the electromagnetic waves traveling in the y direction with polarization vectors $\hat{e}_1$ and $\hat{e}_2$ and frequencies ω_1 and ω_2.

The transition probability per unit of time and per atom between the 1s and the 2s state is then:

$$W_{\text{Raman}}(\text{1s-2s}) = \frac{|E_1 E_2|^2 e^4 a_0^4}{4(2\pi\hbar^2)\Re^3}\frac{1}{9}|D|^2\delta(\Delta\nu), \qquad (2.2)$$

where a_0 is the Bohr radius; $\Re = 3.29 \cdot 10^{15}$ sec^{-1} is the Rydberg frequency and

$$D = \frac{3}{2} \sum (1 + P_{12}) \frac{\hat{e}_1 \cdot \left\langle 2s \left| \frac{r}{a_0} \right| n \right\rangle \left\langle n \left| \frac{r}{a_0} \right| 1s \right\rangle \cdot \hat{e}_2}{\nu_n - \nu 1s - \nu_2} \qquad (2.3)$$

where P_{12} interchanges and polarization of the two photons. The sum in (2.3) is extended only over the discrete spectrum: in fact it has been observed that, in the Hydrogen atom, the continuum spectrum gives a negligible contribution in the r-gauge[6]. denotes the dimensionless frequencies defined as:

$$\omega_i = 2\pi \Re \nu_i, \quad \nu_{1s} = \frac{E_{1s}}{2\pi \hbar \Re} = -Z^2, \quad \nu_{2s} = \frac{E_{2s}}{2\pi \hbar \Re} = -\frac{Z^2}{4}, \quad \nu_n = -\frac{Z^2}{n^2}, \quad (2.4)$$

where ν_n are the frequencies of the intermediate states $|n\rangle$. The argument of the energy-conserving δ function for a Raman process reads

$$\Delta \nu = \nu_{2s} - \nu_{1s} + \nu_1 - \nu_2, \qquad (2.5)$$

where ν_1 is the frequency of the emitted photon and ν_2 is the frequency of the adsorbed photon.

Calculating the integrals of (2.3) for every polarization direction of emitted light[7]

$$\sum_m \left\langle \left| \frac{r}{a_0} \right| n'(l+1)m' \right\rangle \left\langle n'(l+1)m' \left| \frac{r}{a_0} \right| n''m'' \right\rangle$$

$$= [(l+1)/(2l+1)] \cdot R_{nl}^{n'(l+1)} R_{n''l}^{n'(l+1)} \qquad (2.6)$$

and considering the δ function (2.5), we obtain:

$$D = \lim_{n \to \infty} \frac{1}{2} \left[\left(\frac{Z^2}{4} - \frac{Z^2}{n^2} - \nu_1 \right)^{-1} + \left(Z^2 - \frac{Z^2}{n^2} - \nu_1 \right)^{-1} \right] \cdot R_{1s}^{nl} R_{2s}^{nl}. \qquad (2.7)$$

This expression refer to photons polarized in the same direction, for polarization angle θ between the two photons, expression (2.7) should be multiplied by $\cos \theta$.

Note that the result do not change if we exchange the two photon involved.

The evaluation of the (2.7) is performed numerically and the matrix elements are calculated from L. I. Shiff[8] eq. (16.24) and (16.22) for the discrete spectrum.

3. RESULTS

The calculations was performed with the convergence condition $|D_{N+1} - D_N|/|D_N| < 10^5$.

Following graphs we plotted eq. (2.7) for $1 < Z < 25$ and $0 < n < 1000$ (Figure 2(a)), $26 < Z < 60$ and $100 < n < 1100$ (Figure 2(b)), $61 < Z < 92$ and $500 < n < 1500$ (Figure 2(c)) and

$$D \wedge 2 = D \cdot D \qquad (3.1)$$

for $1 < Z < 92$ and $0 < n < 10000$ (Figure 1). Each graph has been constructed with 10000 frequency values for each Z.

4. DISCUSSION

The results of our calculation will be discussed in the light of what was done earlier by Maquet and Rahman[2] and their conclusions.

Maquet and Rahman had shown that at a suitable frequency, the scattering amplitude becomes zero. The zero persists even when the amplitude of the other diagram is added. Now, this shifts the position of zero, and this result is gauge independent. So, one identifies that at a specific frequency, the system is Raman-transparent. Such a phenomenon, have been essentially seen experimentally by Toschek[9] in 1998, with the sodium atom. The gauge dependence of the amplitudes, however, prevents one from identifying a specific frequency at which the emitted Raman photon appears uniquely from the ground state of the hydrogen atom. Now, the computations have been done for all values of Z from 1 to 92.

The entire result of all the computations is presented in a nutshell on Figure 1, which is the transition probability as a function of the "laser" frequency, which is a continuous variable and the charge number Z, which identifies the original atom and is a discrete number.

From Figure 1, one may identify, that the experiments need to be designed for external electromagnetic sources in the frequency range of 1 and 10000.

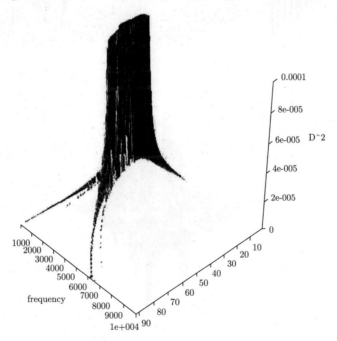

Figure 1. Probability of Raman-like 1s-2s transition for hydrogen-like atoms as a function of Z and the laser frequency ν

It should also be borne in mind that the production of an atom of $Z = 50$, say, may require a combination of intense lasers, so ultimately al but one of the electrons are removed. We shall not belabor that point here, except to reiterate that the Raman-like processes being calculated here, necessities experimental capabilities of producing single electron systems from helium to uranium with a variety of methods and then to go on to perform the Raman-Like experiment. The overall systematics of these processes, i.e. the relevant transition probabilities are summarized in Figure 1.

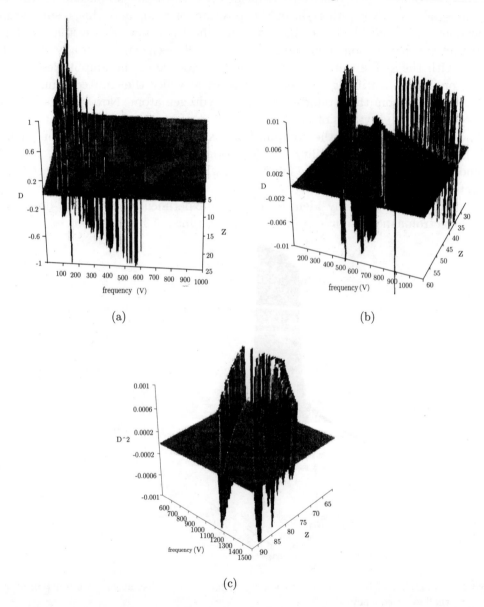

(a)

(b)

(c)

Figure 2. The amplitudes from which the probabilities have been calculated. Note that they are all real (a) amplitudes from $Z = 1$, to $Z = 25$, (b) amplitudes from $Z = 26$, to $Z = 60$, (c) amplitudes from $Z = 61$, to $Z = 92$

In Figure 2, with three figures, the result of all Z, are shown in a larger scale. In Figure 2(a), Z varies between one to 25, in Figure 2(b), it varies between 26 to 60 and in Figure 2(c), the rest of the results up to $Z = 92$ are shown.

5. CONCLUSION

These are the preliminary results for $Z \neq 1$ hydrogenic atoms in external field for Raman-like processes. Details of these calculations, their implications, experimental prospects and relativistic corrections will be published elsewhere[10].

REFERENCES

1. N. K. Rahman, The Hydrogen Atom, G. F. Bassani, M. Inguscio and T. W. Hansh, Springer, Berlin, 1989, 274.
2. A. Maquet and N. K. Rahman, J. Physique, 48 (1987), 1247.
3. C.T. Munger and H. Gould, Phys. Rev. Lett., 57, 23 (1986) 2927.
4. D. Wineland, P. Ekstrom and H. Dehmelt, Phys. Rev. Lett., 31 (1973), 1279.
5. J. Alonso, K. Blaum, S. Djekic, H.-J.Kluge, W. Quint, B. Schabinger, S. Stahl, J. Verdú, M. Vogel and G.Werth, Review of Scientific Instruments, 77 (2006), 03A901.
6. F. Bassani, J. J. Forney and A. Quattropani, Phys. Rev. Lett., 39, 17 (1977), 1070.
7. H. A. Bethe and E. E. Salpeter, Quantum Mechanics of One- and Two-Electron Atoms, Sprinter, Berlin, 1957.
8. L. I. Shiff, Quantum Mechanics, McGraw-Hill, 1955.
9. V. M. Baev, K. J. Boller and P. E. Toschek, Opt. Communic., 66 (1988), 225.
10. S. Fortuna and N. Rahman, preprint, to be published.

Atoms and Molecules in Laser and External Fields
Editor: Man Mohan

Molecular Dynamics in Intense Laser Fields

M. Mohan, A. Maan and P. Jha

Department of Physics and Astrophysics, University of Delhi, Delhi 110 007, India

INTRODUCTION

The study of atoms and molecules in strong laser fields [1-7] with ultra-short-pulse duration has been one of the areas of atomic and molecular physics that has attracted considerable interest recently. The definition of laser-field being intense depends upon its application in various situations. In common day life when the laser cuts through a hard material, then it in regarded as intense. However, on the microscopic scale it might be different. It is still possible that the interaction of individual molecules with the same laser is well treated by perturbation theory, and the laser field is therefore, weak. The difference arises from the large difference between our perception of time and the atomic time scale (1 a.u time $\approx 2.4 \times 10^{-17}$ s). Even on atomic scale different meanings may be used, depending on the situation. In case of interaction of laser with two level systems, the Rabi-frequency describes the strength of the coupling. We say the field is strong, if the Rabi-period is shorter than interaction time (usually the duration of the laser pulse) i.e. $(\Omega < \tau_{\text{pulse}})$. In this case, the population is initially pumped to the 'final' level and then back to the initial level, i.e., absorption is followed by stimulated emission. The result is periodic transfer of population or Rabi oscillations [8]. However, if the final state happens to be continuum, the molecule ionizes or dissociate.

At higher intensities, the Rabi-frequency becomes very large and near to the laser frequency.In such a field, the molecular cloud is distorted and free-wave-function becomes dressed wave-function while corresponding molecular energy levels are shifted across distances of the same order as the separation between levels. Under such field one cam optimize laser-molecular interactions with different laser parameters to achieve different important goals. Such goals might, which are normally unavailable due to selection rules,for example, be efficient production of coherent radiation atto-second lasers (high harmonic generation), production of new products from polyatomic molecules and new products from solids are possible.

For still higher field $I \geq 10^{14}$ W/cm^2 the laser field becomes appreciably stronger than the inter-field of the molecule resulting stripping off the valence electrons irrespective to the molecular species. At more stronger laser fields new phenomenon occurs in molecules and clusters e.g. coulomb-explosion resulting source of high speed charge particles which can be used for further applications.

As we go further to intensities $I \geq 10^{18}$ W/cm^2 we reach to the Super-Intense regime where the interaction with the valence electron is dominant, while the interaction with parent ion is more like a perturbation. At such field Relativistic-Correction also become quite important and plays a dominant role.

INTENSE LASER SYSTEMS

To achieve intense laser one has to squeeze large amount of energy in a sufficiently short pulse time. Various types of table-top intense lasers are available these days. Among them solid state lasers like Nd: YAG, Nd: YLF lasers and gas lasers like Kr-F eximer lasers are quite popular. In the high intense regime, the most popular laser is the femtosecond titanium sapphire laser (Ti: Al$_2$O$_3$) in combination with *chirped pulse amplification* (CPA). Tunability in such a system can be achieved with the help of *optical parametric amplifier* (OPA).

Due to the availability of shorter and shorter pulses, the focussing of lasers to higher intensities have become easier. Example of nano-second pulse laser is that of 10 Hz Nd: YAG, laser ($\lambda \simeq 1064$ nm) which produces pulses of approximately $\simeq 10$ ns with energy $\simeq 1$ J corresponding to the pulse power $\simeq 10^8$ W. To achieve the intensity of 10^{14} W/cm^2, from such a laser, this system should be focussed with spherical lense to the area of $A \simeq 10^{-6}$ cm^2. A typical femo-second laser is Kilo-Hertz Ti: Sapphire laser with $\simeq 100fs$ pulses with energy $\simeq 1$ mJ. The pulse power $\simeq 10^{10}$ W which is higher than the previous one. From this, the intensity can be achieved to $I \simeq 10^{15}$ W/cm^2 by focussing with a lense of focal length $f \simeq 300$ mm.

In order to understand the behaviour of molecule in a strong field one has to understand the basic structure of molecule. The simplest molecule is di-atomic molecule having two atoms. The binding of two atoms occurs when they are close and due to the formation of new electronic state. The valence electrons from the atoms experience double (or multiple) set of coulomb interactions which leads them to occupy new electronic states that are of bonding (or anti-bonding) character. For multi-electron molecule e.g NO$_2$, HOCl, it is the highest occupied molecular orbital (HOMO) state that determines the characterstics of the bond. The orbital may be symmetric or anti-symmetric in character. The electronic state give rise to *potential surface* (PES), which in turn is experienced, by the atomic nuclei. In typical diatomic molecule e.g. H$_2$ we have two lowest potential surfaces (bonding and anti-bonding). For poly-atomic molecules the bond is characterized by two quantities; one, by the angle made with bonds between neighbouring atoms and second by the bond length. The electronic state above the lowest HOMO state are the excited states. The lowest unoccupied excited state, is called the lowest unoccupied molecular orbital or LUMO.

The positions of HOMO, LUMO and other excited states play an important role in the interaction with the laser field i.e. it leads to enhanced excitation if the transition frequency between the states ω_s, is in single or multiphoton resonance with the field. If the excited state (e.g LUMO) to which the molecule is excited due to laser field has no potential minimum but decrease with inter-nuclear distance r_e, then the molecule will dissociate on the time scales from few femto-seconds to pico-seconds depending upon the nature of potential surfaces and mass of the fragments.

We next consider the vibration motion (or nuclei) in the nuclear-bond about the equilibrium distance re in the molecule. The potential surface which has a potential well looks like potential well of harmonic or anharmonic oscillator (e.g Morse-Oscillator). Like harmonic-oscillator one can quantize the vibration motion of the nuclei about r_e into ground and excited vibrational modes with energy spacing nearly equal to $\simeq \hbar\omega_o$ where ω_o is the natural frequency of the oscillator. The vibration period T_v varies from 10 fs for a strongly bound system like N_2 to 10's of ps in case of poly-atomic molecule. We may have "strech" modes which modulate the bond length, or "bend" modes which corresponds to the oscillation in the angle of bond about the equilibrium direction with period of bend motion $T_B \sim 100$ fs.

We know that molecule not only vibrate but can also rotate in space about one or more axis. The speed of rotation depends upon the moment of intertia I about the rotational axis and therefore rotational constant $B = h/8\pi^2$ cI in cm^{-1}. The rotational period of the molecule $T_R \simeq 1/2$ Bc varies from $T_R \simeq 100$ fs to 10's of ps. One has to cool the molecules quite enough so that the molecules may not be the thermal distribution of J-states. A short pulse with duration τ_p interacts with the sample on a time-scale very much less than the rotational period (i.e. $\tau_p \ll T_R$) and thus "freezes" the tumbling of the molecules at one instant so that it might be imagined to be an ensemble of randomly aligned molecules.

In the coming paragraphs we will discuss the various multiphoton processes i.e. multiphoton ionization, coulomb-explosion, High-Harmonic Generation, control of alignment or orientation of an ensemble of an aligned molecule etc. at an instant of interaction with laser pulse.

MULTIPHOTON FIELD IONIZATION AND COULOMB-EXPLOSION

In the strong field there may be multiphoton or tunnel ionization in multi-electron system. Keldysh [10] introduced a useful parameter, γ, now known as Keldysh parameter to separate non-linear multiphoton processes, either atomic or molecular, into two regimes, the multiphoton and tunneling regimes. The Keldysh parameter γ, is defined as the ratio of atomic (molecular) electronic energy and field induced energies,

$$\gamma = \sqrt{\frac{I_P}{2U_p}},$$

where I_p is the ionization potential, $U_p = e^2E^2/4m\omega^2 = I/I_04\omega^2$ a.u is the ponderomotive energy and $I = I_0 = 1$ a.u. is the laser intensity, corresponding to 3.5×101^{16} W/cm^2.

In case of H atom, $I_p = 13.6$ eV $= 0.5$ a.u, so that for laser intensity with $I = I_0 = 1$ a.u, $\omega = 1$ a.u (with $\lambda = 45.6$ nm), $U_p = 1/4\omega^2 = 0.25$ a.u the Keldysh parameter

$$\gamma = \sqrt{\frac{0.5}{2 \times 0.25}} = 1.$$

Thus for $\gamma < 1$ and for low $\omega < I_p$, where $U_p > I_p$, the ionization rate becomes independent of wavelength and ionization and Higher Harmonic Generation can be described by tunneling mechanism.

For $\gamma > 1$, typical of higher frequencies e.g UV, X-ray, ionization is by multiphoton process and ionization rate is αI^n.

Tunnel Ionization for $\gamma < 1$, can be understood by considering the distortion of Coulomb-potential of the atom (or molecule) by strong laser field. The height of the potential is raised when the field becomes large and the potential is lowered when the field becomes low. The bound electron thus in the presence of field feels barrier of finite width through which there is a finite probability of quantum tunneling. Delone and Krainov [40] has given the formula for the quantum tunneling in atom. For strong near infra-red laser field, the tunnel ionization dominate over the multiphoton ionization rate by many order of magnitude.

Multiphoton Ionization in atoms and molecules have been done by several authors [15, 16], A comprehensive study of one, two and four photon ionization of H_2 at the Kr-F laser has been recently studied by Burke group [17]. They have combined R-matrix theory of scattering part with Floquet theory for radiative part as had been employed successfully by Mohan Milfield and Wyatt [18] in laser induced chemical reactions.

Coulomb-Explosion. If the ionization of several electrons in a molecule is rapid due to interaction with laser field, the *coulomb-explosion* occurs. The remaining nuclei of the molecule feels massive repulsion force.

The potential energy stored in the system due to coulomb repulsion between the positively charged nuclei is rapidly released as kinetic energy shared between the two nuclei that fly apart [24, 25]. For a pair of bare nuclei of diatomic molecule like N_2 with $r_e \simeq 10^{-10}$ m, the coulomb potential energy $\simeq 50$ eV. This energy is equally shared between the two nuclei and the acquired velocity 2×10^5 ms^{-1}. In case of polyatomic molecules and clusters the fragment energy due to coulomb-explosion may go up to 100's of keV range [26]. By noting the angular distribution and energy of the fragment one can find the architecture of the molecule, known as Coulomb-explosion Imaging [27].

LASER-INDUCED ORIENTATION AND ALIGNMENT IN MOLECULES

The study of atoms under the impact of intense laser field leads to various multi-photon processes like ATI (*Above Threshold Ionization*) and HHG (*High-Harmonic Generation*). However, if we go little towards complexity i.e. in case of molecules and specifically to polar molecules with large permanent dipole moment, which when subjected to intense laser field, shows an interesting phenomenons called as 'Orientation or Alignment Effect'.

The term 'Orientation' has its significance in the fact that most of the chemical reactions are sensitive to the relative orientation of the reactants.So, it is quite desirable to develop a strategy for orienting the molecules in a particular way and controlled manner.

The Orientation of the molecules can be carried out in several ways using static or time-dependent electromagnetic fields, however the HCP (*half-cycle pulses*) of the form given by $f(t)$ [19],

$$f(t) = \begin{cases} \sum_{j=0}^{N=1} \sin^2[\Pi(t-j\tau)/T] & 0 < t - j\tau < T \\ 0 & \text{otherwise} \end{cases}$$

is of particular interest as they form a train of pulses all pointing in the same direction, not flipping the sign of the field for alternative half-cycle and thus produces better orientation Here $j = N - 1$ and τ is time delay between two sequential HCP's. The rotational period $T_{\rm rot} \simeq h/2{\rm B}$. Now, to explain the term Orientation in a more elaborate and analytical way, let us assume that the electric field takes the form

$$\vec{E}(t) = E_0 f(t) \cos(\omega_o t)$$

where

E_o = electric field amplitude, $f(t)$ = pulse envelope, ω_o = angular frequency.

Thus, the Hamiltonian will include in it the terms involving interaction of the dipole moment (μ_0) of the molecule and the electric field E; and also the interaction of polarizability components ($\alpha\perp$ and a_{Pi}) charactistic of a polar molecule, with the electric field. Thus, the generic Hamiltonian will take the form as given[21, 22]

$$H = BJ^2 + V_\mu(\theta, t) + V_{\rm pol}(\theta, t)$$

where

B = rotational constant of the molecule,

J^2 = squared angular momentum operator,

θ = angle between electric field and the molecular axis.

The term

$$V_\mu(\theta, t) = -\mu_o E(t) \cos\theta$$

where μ_0 = permanent dipole moment of the molecule along the inter nuclear axis and the term

$$V_{pol}(\theta, t) = -\frac{1}{2}[\Delta\alpha \cos^2(\theta, t) + \alpha_\perp]E^2(t)$$

where

$$\Delta\alpha = \alpha\,\mathrm{II} - a_\perp.$$

Thus the resulting TDSE (time-dependent Schrodinger equation)becomes

$$i\hbar\frac{d\Psi(\theta,\phi;t)}{dt} = H(t)\Psi(\theta,\phi;t).$$

The above TDSE can be solved numerically using either the Split operator technique[23]on a discrete grid or a basis set expansion of the wavefunction in terms of spherical harmonics propagated with a Fourth-Runge-Kutta Scheme, with initial state taken as the ground rotational (isotropic) state $J = M = 0$ offering the advantage of rotational excitation only to high J. The measure of orientation and alignment can then be obtained as [21, 22]

$$\langle\cos^n(\theta)\rangle(t) = \int_0^{2\pi}\int_0^{\pi}|\Psi(\theta,\phi;t)|^2\cos^n(\theta)\sin(\theta)d\theta d\phi$$

the expectation value of $\cos^n(\theta)$. Here in the above equation $n = 1, 2$ refers to orientation and alignment parameters or factors respectively. It can be seen that the value of the Orientation parameter i.e. $\langle\cos(\theta)\rangle$ varies within the interval $[-1, 1]$ and that of alignment parameter i.e. $\langle\cos^2(\theta)\langle$ varies between 0 and 1. The extremal value of $\langle\cos^n(\theta)\rangle$ defines the extent to which the orientation or alignment is being achieved

HHG (HIGH-HARMONIC GENERATION) IN ATOMS AND MOLECULES

The process of emission of photons started with the Einstein explaination of the photoelectric effect which is a "scarce photon" theory. However, with the advent of lasers, it became possible to send thousands or even millions of photons through a typical cross-section.Thus, lasers took away the "scarce photon" theory to "multiphoton" theory. The High-Harmonic is one of the results of the multiphoton theory. It says that when an quantum system is subjected to an intense laser field, the response of the system to the external field becomes highly non-linear [28] as the laser intensity rises to the order of 10^{13} W/cm^2 and higher. The result is the radiation of photons by the system at frequencies of multiples of laser fundamental frequency which is normally called as "Harmonic generation(HG)".

The study of HHG (*High-Harmonic Generation*) with HRL's (*Hyper Raman Lines*) is important in the sense that it creates the possibility of obtaining high-frequency light. The process can be explained using an equation as explained below:

Suppose a target A (i.e. atom) absorbs q photons (i.e. γ) then as a result of the process, we get the same atom and a photon γ' whose energy is q times the energy of the incident photon [29]

$$A + q\gamma(\hbar\omega_L) \rightarrow A + \gamma'(q\hbar\omega_L).$$

To explain the process analytically, we consider the wave function of the form [30] i.e.

$$\Psi(t) = \langle c_i(t)|\phi_i\rangle,$$

where i is the number of states involved in the system and c_i's are the corresponding amplitudes.

The Hamiltonian of the system consists of two type of terms: diagonal and the off-diagonal terms including in it the energy and the interaction terms respectively [30].

$$H(t) = \begin{bmatrix} -\omega_o/2 & \Omega_o \sin(\omega_L t) & \dots & \dots & \dots \\ \Omega_o \sin(\omega_L t) & \omega_o/2 & \dots & \dots & \dots \\ \dots & \dots & \dots & \dots & \dots \\ \dots & \dots & \dots & \dots & \dots \\ \dots & \dots & \dots & \dots & \dots \end{bmatrix},$$

where ω_L is the laser frequency Ω_o is the rabi frequency which is given by

$$\Omega_o = -\mu \cdot E_o, \tag{1}$$

where μ is the electric dipole transition matrix element between two states of the system. Thus, the Schrodinger equation(in atomic units) takes the form

$$d(t) = \langle \Psi(t)|d|\Psi(t)\rangle,$$

$$i\frac{d}{dt}\begin{pmatrix} c_1(t) \\ c_2(t) \\ \dots \\ \dots \\ c_i(t) \end{pmatrix} = \begin{pmatrix} -\omega_o/2 & \Omega \sin(\omega_L t) & \dots & \dots & \dots \\ \Omega \sin \omega_L t & \omega_o/2 & \dots & \dots & \dots \\ \dots & \dots & \dots & \dots & \dots \\ \dots & \dots & \dots & \dots & \dots \\ \dots & \dots & \dots & \dots & \dots \end{pmatrix}\begin{pmatrix} c_1(t) \\ c_2(t) \\ \dots \\ \dots \\ c_i(t) \end{pmatrix}$$

and then the time-dependent dipole moment becomes where

$$\vec{d} = -\vec{E(t)} \cdot \vec{r}$$

where

$$E(t) = \text{Electric Field}$$

So, its fourier transform of above equation takes the form

$$d(w) = \left| \int dt e^{-iwt} d(t) \right|^2.$$

The plot of $\log(d(\omega))$ with the harmonic order (i.e. ω/ω_o) gives the required harmonic spectra and thus explains Harmonic generation and its characteristics like plateau,cutoff etc.

With the possibility of aligning the molecules Velotter *et al.* [31], and Corkum *et al.* [33] found the modification in HHG in CS_2, H_2 and O_2 molecules.Experimental studies in HHG was done for the first time by Chin's group [34].Recently experiments on looking at the symmetry effects (e.g. O_2 versus N_2) has been done by Lin Group [32] where they have shown that for π_g bonded systems such as O_2 the harmonic cutoff is extended beyond the limit found in the partner atom (Xe), apparently due to suppression of ionizaton leading to an elevated ionization saturation intensity in this molecule.

Ellipticity of laser field is found to play a very sensitive role in the yield of HHG in atoms and molecules [35, 36] as even a small value of elllipticity suppress the recollision of the electron wavepacket with the core.Thus ellipticity dependent HHG studiescan give directly the information about the electron dynamics. Theoritical work on HHG in H_2^+ and H_2 molecules have been done for the first time by Bandrauk and Co-workers [37,39]. other groups active in this area is of Becker *et al.* [38], and Knight [40]. Knight *et al.* [42] have adopted the numerical technique and have shown that the amplitude and phase of single molecule HHG response in both H_2^+ and H_2 varies as the angle between the molecular axis and the laser polarization direction θ. In their work dipole amplitude was found which shows a minimum at some intermediate angles θ (corresponding to which there is a substantial jump in phase of the dipole by around θ) [40] and is due to interference in harmonic emission between two atomic centers in the molecule [41].

CONCLUSION

We have discussed several intense-laser-molecular interaction phenomenon due to development of laser technology where one can control a light pulse on the time scale of atomic and molecular processes. Another important goal of research in this field is to channel light into a particular chemical bond and catalyze the synthesis of entirely new compounds at room temperature and normal atmospheric pressure.In this direction Mohan et.al have predicted the possibility and JILA (U.S.A) group have demonstrated the ability to suppress or enhance specific vibrational modes in benzene and other polyatomic molecules by changing the temporal shape of a light pulse.Such type of theoretical and experimental work will help to understand how to create complex vibrations in molecules and eventually to control chemical reactions.

REFERENCES

1. C. Berry and M. Perry, Corkum, Private Communication G. Mourou, Phys. Today, 22 (1), 1998.

2. M. Mohan, Current Development in Atomic, Molecular, Chemical Physics with Applications, Kluwer Plenum Press, 2002.

3. A. D. Bandrauk (editor), Molecules in Laser Fields, Marcel Dekker, New York, 1994; A. D. Bandrauk, Y. Fujimura and R. J. Gordon (editors), Laser Control and Manupulations of Molecules, 2002, ACS Symposium Series, 821, Oxford University Press.

4. I. N. Levine, Quantum Chemistry, 3rd edition, Pearson Education Singapore Pvt. Ltd., 2003.

5. M. Gavrila, Atoms in Intense Laser Fields, Academic Press, Orland, 1992.

6. M. Protopapas, C. H. Keitel and P. L. Knight, Rep. Prog. Phys., 60 (1997), 387.

7. J. H. Posthumns, Rep. Prog. Phys., 67 (2004), 623.

8. M. Padgent and L. Allen, Phys. World, 10 (1997), 35.

9. C. Roos and T. Zeiger, Phys. Rev. Lett., 83 (1999), 4713.

10. L. Keldysh, Sovt. Phys. JETP 20 (1965), 1307.

11. M. Mohan, Invited talk in National Conf.

12. R. Kundliya and M. Mohan, Phys. Lett. A, 291 (2001), 22.

13. N. Singhal, V. Prasad and M. Mohan, Eur. J. Phys. D, 21 (2002), 293.

14. K. Batra, R. Kundliya and M. Mohan, Pramana J. Phys., 62 (2004), 31.

15. E. Cornier and P. Lambropolous, J. Phys. B, 30 (1997), L17.

16. R. Kundliya, V. Prasad and M. Mohan, J. Phys. B, 33 (2000), 3263, R. Kundliya and M. Mohan, Phys. Rev. A, 64 (2001), 043404.

17. P. G. Burke, J. Colgan, D. H. Glass and K. Higgins, J. Phys. B: At. Mol. Opt. Phys., 33 (2000), 143.

18. M. Mohan, K. F. Milfeld and R. E. Wyatt, Chem. Phys. Lett., 99 (1983), 411.

19. S. Hu and L. Collins, Phys. Rev. A, 69 (2004), 041402.

20. A. Matos and J. Berakder, Phys. Rev. A, 68 (2003), 063411.

21. O. Atabek *et al.*, J. Phys. B, 36 (2003), 4667.

22. A. Yedder *et al.*, Phys. Rev. A, 66 (2002), 063401.

23. C. E. Dateo and M. Horia, J. Chem. Phys., 95 (1991), 7392.

24. L. J. Frasinski, K. Codling and P. A. Hatherly, Science, 256 (1989), 1029.

25. C. Cornaggia, M. Schmidt and D. Normand, J. Phys. B, 27 (1994), L123.

26. M. Lezius, S. Dobosz, D. Normand and M. Schmidt, Phys. Rev. Lett., 80 (1998), 261.

27. J. Levin, H. Feldman, A. Baer, D. Ben-Hamu, O. Heber, D. Zalfman and Z. Vager, Phys. Rev. Lett., 81 (1998), 3347.

28. T. Zuo, S. Chelkowski and A. D. Bandrauk, Phys. Rev. A, 48 (1993), 3837.

29. J. L. Krause, K. J. Schafer and K. C. Kulander, Phys. Rev. A, 45 (1992), 4998.

30. C. Liu, S. Gong, R. Li and Z. Xu, Phys. Rev. A, 69 (2004), 023406.

31. R. Velotta, N. Hay, M. R. Mason, M. Castillejo and J. P. Marangos, Phys. Rev. Lett., 87 (2001), 183901.

32. B. Shan, X. M. Tong, Z. Zhao, Z. Chang and C. D. Lin, Phys. Rev. A, 66 (2002), 061401 (R).

33. P. Corkum *et al.*, High-Harmonics from Aligned Molecules, Post Deadline Paper, in Proceedings of Ultra-Fast 2003, Vienna.

34. Y. Liang, S. Augst, S. L. Chin, Y. Beaudion and M. Chaker, J. Phys. B, 27 (1994), 5119.

35. N. Hay, R. de Nalda, T. Halfmann, K.J. Mendham, M. B. Mason, M. Castillejo and J. P. Marangos, Eur. Phys. J. D., 14 (2001), 231.

36. A. Flettner, J. Konig, M. B. Mason, T. Pfeifer, U. Weichmann, R. Durer and G. Gerber, Eur. Phys. Journal D, 21 (2002), 115.

37. H. Yu and A. D. Bandrauk, J. Chem. Phys., 102 (1995), 1257.

38. R. Kopold, W. Becker and M. Kleber, Phys. Rev. A, 58 (1998), 4022.

39. T. Zuo an A. D. Bandrauk, Phys. Rev. A, 52 (1995), R2511.

40. M. Lein, N. Hay, R. Velotta, J. P. Marangos and P. L. Knight, Phys. Rev. Lett., 88 (2002), 18903.

41. M. Lein, N. Hay, R. Velotta, J. P. Marangos and P. L. Knight, Phys. Rev. A, 66 (2002), 023805.

42. P. Corkum, Private communication.

Atoms and Molecules in Laser and External Fields
Editor: Man Mohan

Spectroscopic Study of Fano Interaction in Laser-Etched Silicon Nanostructures

Rajesh Kumar, A. K. Shukla and H. S. Mavi

Department of Physics, Indian Institute of Technology, Hauz Khas, New Delhi 110 016, India

INTRODUCTION

The semiconductor nanostructures (NSs) have been investigated in recent years because of their immense use in electronic and opto-electronic devices [1-3]. Electronic and optical properties of semiconductor nanostructures are strongly affected by quantum confinement of carriers due to the reduced dimensionality of these systems. When the motion of carriers is limited in more than one dimension, quantum confinement results in a strong increase of the optical photoluminescence (PL) energies. One expects that other optical and electronic properties will also be affected as well. In heavily doped n-type silicon (Si) [4, 5], p-type Si [6, 7] and ion-implanted Si [8], the effect of interband electronic excitations on the Raman spectra has been studied in term of the electron-phonon coupling.

The interaction of electronic transitions with discrete optical phonons results in Fano interference. As a result, the Lorentzian Raman line-shape changes to a typical asymmetric Fano-type line-shape. The Fano interference has also been seen in intersubband transitions in the case of quantum wells [9]. In such semiconductor systems, the phonon states and the electronic states involved in the Raman scattering are discrete quantized levels because of confinement. Optical transitions between these electronic levels have been investigated in the PL and electroluminescence spectra of semiconductor NSs [10, 11]. In Si NSs of size 2-4 nm, the PL spectra usually show the presence of discrete electronic levels due to quantum effect. It may give rise to the interference of Raman active phonons with photo-excited carriers in electronic levels in the Raman scattering.

It has been reported that heavy doping of 4×10^{19} cm^{-3} is required to observe the Fano interference in n-type crystalline Si (c-Si) [12]. In suitably designed semiconductor super lattices, Fano-interference of electronic intrasubband transitions with optical phonons has been studied [13]. In super lattices, the conduction subband can be easily tailored to be wide enough to overlap on optical phonons. Fano interference involving the photo-excited carriers and the phonons can be seen if an excitation laser power density of the order of $10^6 - 10^7$ W/cm^2 is used [14, 15] for a c-Si sample. However, these properties are expected to alter in highly confined systems such as Si quantum dots or wires. Recently, Gupta *et al.* [15] have shown laser-induced Fano resonance in Si nanowires of dimension 5-15 nm prepared by pulsed laser vaporization. Laser power densities in our experiment are lower than those reported earlier [15], because the sizes of nanocrystallites are around 5 nm in our sample. We believe that a substantial reduction in the laser power density may be due to quantum confinements.

The Raman spectra from nanocrystalline Si have been extensively studied [16-23]. The broadening and downshift of the optical phonon mode is due to phonon confinement in Si nanostructures. According to the phonon confinement model [24, 25], relaxation in the k-vector selection rule for the excitation of the Raman active optical phonons gives rise to a broad Raman spectrum. Sizes of the crystallites can be estimated using this model. In our experiments, one can observe an asymmetric broadening and red shift in the Raman peak of laser-etched NSs in comparison to that of the c-Si. In addition, one can see changes in asymmetry ratio of Raman mode on increasing laser power density for laser-etched sample. The asymmetry ratio is defined here as Γ_a/Γ_b, where Γ_a and are half widths on the low- and high-energy side of the maximum. Raman spectra for crystalline and ion-implanted Si samples are also discussed here and no such dependence of Raman mode of the sample on laser power density is seen. By this comparative study, one can invoke that Fano interference takes place at comparatively smaller laser power density in Si NS than that for the bulk c-Si.

Extensive studies of photo-excited carriers on Raman spectra of Si NS, ion-implanted Si and c-Si are presented here. The Si NSs are fabricated by *laser-induced etching* (LIE) [26]. Surface morphology of laser-etched sample was studied by *atomic force microscope* (AFM). Most probable dimension of nanocrystallites is determined by phenomenological phonon confinement model including Fano interference. The PL spectrum from Si NS shows multiple peaks.

EXPERIMENTAL DETAILS

Sample A is a commercially available n-type Si wafer, having resistivity 10Ω-cm. Sample B is prepared by LIE [26] of sample A. The LIE was done by immersing the wafer into 48% HF acid in a plastic container and then focusing a 100 mW argon-ion laser beam ($E_{ex} = 2.41$ eV) to a circular spot of 120 mm diameter on the wafer. Sample was etched in this way for 30 minutes and was subsequently rinsed with ethyl alcohol. It was dried in air before recording the Raman and PL spectra. Sample C was prepared from sample A by arsenic ion-implantation with arsenic at 100 keV with fluence rate of 2.5×10^{16} ions/cm^2 and was subsequently annealed with Nd: YAG laser with energy density of 1.4 J/cm^2. Raman and PL spectra were recorded by employing a spectroscopic system consisting of a SPEX-1403 doublemonochromator, a HAMAMATSU (R943-2) photo multiplier tube, an amplifier-discriminator assembly, a photon counter, a computer and an argon-ion laser (COHERENT, INNOVA 90). The Raman spectra were recorded using an excitation photon energy 2.54 eV of the argon-ion laser at four power densities ranging from 0.22 to 1.76 kW/cm^2. The surface morphology of sample B containing nanoparticles was studied by AFM (Digital Instruments NanoScope), in contact mode.

EXPERIMENTAL RESULTS

Figure 1 shows the surface morphology of laser-etched sample (Sample B) by AFM. The image is the inside AFM image taken from pore walls. It shows approximately 30 dots in a scan area 500 nm 500 nm, which gives a dot density 1.21012 cm^{-2}. Figure 1 shows that the dots are distributed randomly, having a width of approximately 5 nm and height around 3 nm. The nanocrystallite size can be controlled by the laser power density and irradiation time during etching.

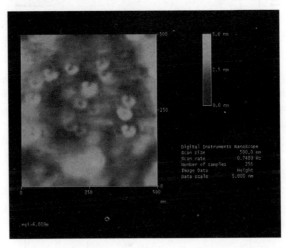

Figure 1. High resolution AFM image, showing the surface morphology of one of the pore walls of sample B

Figures 2(a)-(c) show the Raman spectra from sample B recorded using excitation photon energy 2.54 eV with different laser power densities. Raman active optical phonon mode, which is observed at 520 cm^{-1} for c-Si (sample A), shifts towards lower wavenumber (518 cm-1) and is found to be asymmetric with FWHM of 13 cm^{-1} for laser-etched sample (sample B). Phonon softening of the Raman peak increases with increasing laser power density in Figure 2. The increase in asymmetry ratio with increasing laser power density is also observed in Figure 2.

Figure 3(a) shows the Raman spectrum of sample C (Arsenic ion-implanted Si), recorded using an excitation photon energy of 2.54 eV with power density of 0.22 kW/cm^2. The Raman spectrum of c-Si is also shown in Figure 3(b) for comparison. Asymmetry in the Raman line-shape of sample C can also be seen. The Raman spectra of sample C are also recorded with higher excitation laser power densities. No changes in the Raman peak position, FWHM and asymmetry are observed with increasing laser power density for sample C. It rules out photo-excited carriers contribution to Fano effect in the sample C.

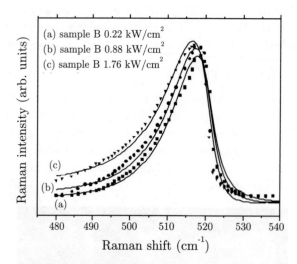

Figure 2. Comparison of the experimental and the theoretical Raman spectra of sample B using Eq. (2). Solid lines indicate the calculated Raman spectra and the experimental data are plotted as discrete points

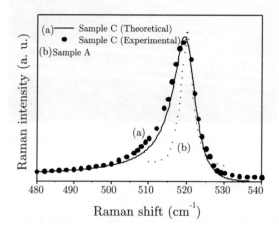

Figure 3. Raman spectra of (a) arsenic ion-implanted silicon, taken at laser power density 0.22 kW/cm2 and (b) crystalline silicon. Discrete points show experimental data and line shows the theoretical fitting using Eq. (1)

In order to investigate the electronic states of the Si NSs, the PL spectroscopy on laser-etched sample (sample B) are also performed. Figure 4 displays the PL spectrum recorded from sample B using an excitation photon energy of 2.54 eV. Multiple PL peaks from Si NS can be seen at 2.1 and 2.0 eV. One hump-like structure is also seen around 1.8 eV. Because of confinement effect, all peaks of PL are observed in the visible region. Multiple PL peaks are similar to what has been reported earlier [10].

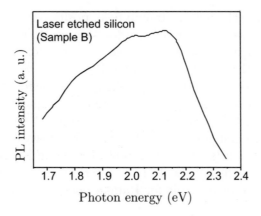

Figure 4. Photoluminescence (PL) spectrum of laser etched silicon (sample B), showing multiple peaks

They have attributed the multiple peaks to the discrete electronic levels of Si nanocrystals. Recently, Ilya Sychugov *et al.* [27] have studied single dot optical spectroscopy of Si nanocrystals. They have shown that PL width of 150 meV at room temperature changes to 25 meV at 80 K. Furthermore, it splits into two electronic states. It is emphasized that position of these states may vary with size of nanocrystals. Since there is a distribution of sizes in our laser etched sample, these electronic states may form continuum that is seen in our broad PL spectrum. Electronic Raman scattering may take place due to participation of photo-excited carriers, which undergo transitions within the continuum of electronic states. Such electronic Raman scattering may interfere with the Raman optical-phonons that are observed for nanocrystallites of various sizes. Fano interaction, which is due to interference between the electronic Raman scattering and phonon Raman scattering, is seen in our laser-etched sample. Transitions of photo-excited carriers within the continuum cause electronic Raman scattering. PL spectrum emphasizes that electronic levels are present and electrons in these states are available for Fano interference.

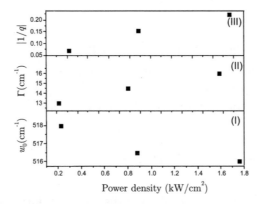

Figure 5. Excitation laser power density dependence of (I) Raman peak position (ω_0), (II) FWHM (Γ), and (III) Fano asymmetry parameter ($|1/q|$) for sample *B*

Figure 5(I) shows the changes in Raman peak position from 518 cm^{-1} to 516 cm^{-1} on increasing laser excitation power density for sample B. There are no such changes caused by laser power density on Raman peak position for sample A and sample C. Figure 5(II) shows that the FWHM for sample B increases with increasing laser power density. Figure 5(III) shows the laser power density dependence of asymmetry parameter for sample B. Our experiments show that asymmetry parameter is independent of laser power density for sample C but is dependant for sample B. For sample A, the Raman line-shape is symmetric. On increasing the laser power density on laser-etched Si (sample B), Raman peak shifts towards lower wavenumber side and FWHM increases. This can happen due to two reasons, firstly, due to Fano interference. Secondly, CW laser heating of the surface may also be responsible for Raman peak shift and increase in FWHM.

DISCUSSION

Fano interference is usually seen in Si if interaction between discrete level phonons and continuum of electronic states take place. Fano interference [28] can be written by asymmetric Fano resonance line-shape, which is given by:

$$I(\omega) = \frac{(\varepsilon + q)^2}{1 + \varepsilon^2}, \tag{1}$$

where, $\varepsilon = \frac{\omega - \omega_0}{\Gamma/2}$. The is Fano asymmetry parameter and is the renormalized resonance frequency with as FWHM. The magnitude of $1/q$ is a measure of asymmetry. Higher $|1/q|$ means larger asymmetry. The value of q can be +ve or $-$ve, depending upon whether the asymmetry is observed towards higher or lower wavenumber side of the Raman peak position respectively. If discrete phonon states, responsible for the typical lorentzian Raman line-shape interfere with electronic Raman involving continuum of electronic states, asymmetric Raman line-shapes are observed due to Fano interference [9]. Sample B has electronic states, which have been revealed in the multiple PL peaks. Thus, increase in the asymmetry ratio of Raman line-shape on increasing laser power density can be explained in terms of Fano interaction of photo-excited carriers with phonons.

For sample B containing nanoparticles, the Fano line-shape described by Eq. (1) cannot fit our experimentally observed Raman line shape. Therefore, we have used modified Eq. (1), by introducing a weighing function $e^{\frac{-k^2 L^2}{4a^2}}$ in the k-space in the Fano-Raman line-shape. Integration of such function for a given size of nanocrystallite of size over the relaxed wavevector provides the desired Fano line-shape of Si NSs, which can be written as:

$$I(\omega) = \int_0^1 \left[\frac{(\varepsilon + q)^2}{1 + \varepsilon^2} \right] \cdot e^{\frac{-k^2 L^2}{4a^2}} d^2 k, \tag{2}$$

where $\varepsilon = \frac{\omega - \omega(k)}{\Gamma/2}$. The $\omega(k)$ is the phonon dispersion relation of the optic phonon of c- Si and is given by $\omega(k) = \sqrt{A + B \cos \frac{\pi k}{2}}$ with $A = 171400$ cm^{-2}, and $B = 100000$

cm^{-2}. The Γ is the line width and crystallite size is L. Lattice constant, 'a' of Si is 54.3 A. The experimental data (discrete points) in Figures 2(a)-(c) have been fitted using Eq. (2), considering two-dimensional confinement for porous Si [26]. For the lowest laser power density of 0.22 kW/cm^2, the value of nanocrystallite size $L \sim 4$ nm is obtained by keeping $|1/q|$ at minimum because Fano interaction is almost negligible at very low power density. On increasing laser power density from 0.22 kW/cm^2 to 1.76 kW/cm^2, asymmetry parameter of Raman line-shape increases as shown in Figure 2. A reasonably good theoretical fit of Raman line-shape (continuous lines) as shown in Figure 2, is obtained for different values of q keeping nanocrystallite size same. The values of $|1/q|$ corresponding to best fitting are plotted in Figure 5(III).

Figures 2(a)-(c) show a good fitting between experimental data and theoretical curve using Eq. (2). In Eq. (2), the exponential term takes care of the broadening in the Raman line-shape due to confinement of phonons in nanocrystals and provides the information about the nanocrystallite size. The parameter in the square bracket is asymmetric line-shape function taking care of Fano interference. The electronic states in Si NS have been confirmed from its PL spectrum (Figure 4), which show multiple peaks. Similar multiple peak structures have been reported earlier [10] for Si quantum dots.

Furthermore, PL spectra did not show any peak between 1.7 eV and 2.4 eV in sample A and C, i.e. no electronic states are available. In the absence of electronic states, Fano effect due to photo-excited carriers is absent and laser power density dependence is not seen for sample A and sample C in Figure 5. Therefore, the asymmetry in Raman line-shape is attributed to the Fano interference between discrete phonon line and photo-excited carriers in electronic levels of nanocrystallites.

For ion-implanted sample (sample C), the experimental data (solid circles) in Figure 3(a) have been fitted with Eq. (1). We observe that there is a good agreement between experimental and theoretical curves using Eq. (1), supporting the occurrence of Fano resonance between the phonon (discrete) states and the inter-conduction band electronic (continuum) excitations in sample C [6]. The continuum of electronic levels is present due to high carrier concentration resulting from implantation. On increasing the excitation laser power density, there are no changes in Fano interference. Therefore, laser power density dependence of 0, or $|1/q|$ is not observed. On contrary, the dependence with laser power density for sample B is observed.

There is a possibility that the increase in the asymmetry parameter with increasing laser power density is due to laser heating of the sample B. If this is the case, the same effect should also be observed in the Raman spectra of bulk c-Si and ion-implanted Si samples. Since the absorption coefficient of porous silicon quantum structures is less than that of bulk c-Si [29], higher laser power density is required to show changes in the Raman line-shape of sample B due to heating as compared to bulk c-Si. In our case, the laser power density was not high enough to do appreciable heating, which may affect the Raman spectra by anharmonic effect. Hence, speculation about attributing the observed increase in asymmetry parameter of the Raman line-shape of sample B to the laser heating effect can be discarded. It is suggested that the photo-excited electrons present in electronic levels (shown by multiple PL peaks) in Si NSs interfere with phonons and results in asymmetric Fano Raman line-shape.

CONCLUSIONS

Laser power density dependent Raman experiments on Si NSs indicate that the Raman line-shape becomes more asymmetric, wide and shifts to lower wavenumber for smaller laser power density in comparison to bulk c-Si and ion-implanted Si. Quantum confinement of photo-excited carriers leads to Fano effect for small laser power density in Si NSs. The laser power density dependence of the Raman line-shape is explained on the basis of Fano interference effect for nanocrystals. Our experiments show that the electron-phonon coupling can be achieved without heavy doping of semiconductor and without using higher laser power densities for Si NSs. The Fano interference between discrete phonon levels and electronic states in Si NSs is observed. PL spectrum of Si NSs indicates that the electronic levels are present in nano-structured Si. Photo-excited carriers in the electronic levels are responsible for Fano interference in Si NSs prepared by laser-induced etching. On the whole, continuum of electronic states and quantum confinement of photo-excited carriers will be useful for the photonic devices and solar cell made of Si NSs. Photo-excited carriers may be responsible for the nonlinear rise of refractive index that has been reported earlier [30-31].

Acknowledgements

Authors acknowledge the financial support from the Department of Science and Technology, Govt. of India under the project "Optical Studies of Self-assembled Quantum Dots of Semiconductors". Authors also acknowledge Prof. V.D. Vankar (IIT-Delhi) for many useful discussions with him. Technical support from Mr. N.C. Nautiyal is also acknowledged.

REFERENCES

1. Y. Cui and C. M. Lieber, Science, 291 (2001), 851.

2. M. H. Huang, S. Mao, H. Feick, H. Yan, Y. Wu, H. Kind, E. Weber, R. Russo and P. Yang, Science, 292 (2001), 1897.

3. X. Duan, Y. Huang, Y. Cui, J. Wang and C. M. Lieber, Nature, 409 (2001), 66.

4. M. Chandrasekhar, J. B. Renucci and M. Cardona, Phys. Rev. B, 17 (1978), 1623.

5. M. Jouanne, R. Beserman, I. Ipatova and A. Subashiev, Solid State Commun., 16 (1975), 1047.

6. F. Cerdeira and M. Cardona, Phys. Rev. B, 5 (1972), 1440.

7. M. Balkanski, K. P. Jain, R. Beserman and M. Jouanne, Phys. Rev. B, 12 (1975), 4328.

8. H. Engstrom and J. B. Bates, J. Appl. Phys., 50 (1979), 2921.

9. K. Jin, S. Pan and G. Yang, Phys. Rev. B, 50 (1994), 8584.

10. X. L. Wu and F. S. Xue, Appl. Phys. Lett., 84 (2004), 2808.

11. K. Mukai, N. Ohtsuka, H. Shoji and M. Sugawara, Appl. Phys. Lett., 68 (1996), 3013.

12. N. H. Nickel, P. Lengsfeld and I. Sieber, Phys. Rev. B, 61 (2000), 15558.

13. S. Pan, Z. Chen, K. Jin, G. Yang, Y. Huang and T. Zhao, Z. Phys. B, 101 (1996), 587.

14. V. Magidson and R. Beserman, Phys. Rev. B, 66 (2002), 195206.

15. R. Gupta, Q. Xiong, C. K. Adu, U. J. Kim and P. C. Eklund, Nano Lett., 3 (2003), 627.

16. N. Nakano, L. Marville and R. Reif, J. Appl. Phys., 72 (1992), 3641.

17. B. Pivac, K. Furic and D. Desnica, J. Appl. Phys., 86 (1999), 4383.

18. B. Li, D. Yu and S. Zhang, Phys. Rev. B, 59 (1999), 1645.

19. P. Mishra and K.P. Jain, Phys. Rev. B, 62 (2000), 14790.

20. R. Wang, G. Zhou, Y. Liu, S. Pan, H. Zhang, D. Yu and Z. Zhang, Phys. Rev. B, 61 (2000), 16827.

21. M. J. Konstantinovic, S. Bersier, X. Wang, M. Hayne, P. Lievens, R. E. Silverans and V. V. Moshchalkov, Phys. Rev. B, 66 (2002), 161311.

22. S. Piscanec, M. Cantoro, A. C. Ferrari, J. A. Zapien, Y. Lifshitz, S. T. Lee, S. Hofmann and J. Robertson, Phys. Rev. B, 68 (2003), 241312.

23. S. Piscanec, A. C. Ferrari, M. Cantoro, S. Hofmann, J. A. Zapien, Y. Lifshitz, S. T. Lee and J. Robertson, Mater. Sci. Eng. C, 23 (2003), 931.

24. H. Richter, Z. P. Wang and L. Ley, Solid State Commun., 39 (1981), 625.

25. I. H. Campbell and P. M. Fauchet, Solid State Commun., 58 (1986), 739.

26. B. G. Rasheed, H. S. Mavi, A. K. Shukla, S. C. Abbi and K. P. Jain, Mater. Sci. Eng. B, 79 (2001), 71.

27. I. Sychugov, R. Juhasz, A. Galeckas, J. Valenta and J. Linnros, Opt. Mater., 27 (2004), 973.

28. U. Fano, Phys. Rev., 124 (1961), 1866.

29. D. Kovalev, G. Polisski, M. Ben-Chorin, J. Diener and F. Koch, J. Appl. Phys., 80 (1996), 5978.

30. S. Vijayalakshmi, M. A. George and H. Grebel, Appl. Phys. Lett., 70 (1997), 708.

31. S. Lettieri, O. Fiore, P. Maddalena, D. Ninno, G. D. Francia and V. La Ferrara, Opt. Commun., 168 (1999), 383.

Atoms and Molecules in Laser and External Fields

Editor: Man Mohan

Application of the Activation Process Model to the Molecules, Positive Molecular Ions, Clusters and Proteins Surrounded of IR Laser Radiation

Anatoly V. Stepanov

National Ozone Monitoring Research and Educational Centre, Byelorussian State University, 7-817 Kurchatov Strret, Minsk 220064, Republic of Belarus

GENERAL MODEL FOUNDATIONS

Two main model foundations have been stated in Ref. [1, 2]: (1) in the elementary activation interaction act, the potential energy of moving atoms changes discretely or in quanta: elementary activation act appears to be a series of quantum subsystems occurring in sequence (these subsystems may also be defined as identical quantum oscillators); (2) in statistical equilibrium with thermal radiation energy exchange between IR radiation and interacting atoms results in discrete translational motion changes of those atoms which absorb and subsequently emit oscillation energy quantum series (Figure 1).

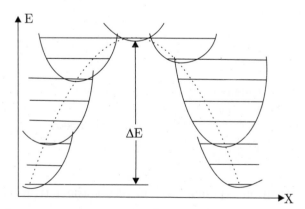

Figure 1. Activation barrier presented in the form of a quantum subsystems sequence

In Ref. [4] the quantum subsystem notion for the non-adiabatic approach has been given: the quantum subsystem is a potential well with its own infinite set of the vibrational levels which is formed because of non-adiabatic interaction of the vibrating nuclei with electronic shell of the molecule. On the one hand, this motion is based on cosequences of the Hehenberg-Kohn-Levy density functional theory [8], on the other hand, it rests on a conditional probability amplitude analysis in wave mechanics [9].

The average energy of elementary activation act for the discretely and translationary moving atoms has been derived from first principles in the form [3]:

$$\bar{\varepsilon} = hv \frac{1 + A[Q^{\xi}(1 - \xi D)] + B[Q^{\eta}(1 - \eta D)] + C[Q^{\varsigma}(1 - \varsigma D)]}{(1 + AQ^{\xi} + BQ^{\eta} + CQ^{\varsigma})(1 - Q)} \cdot Q \tag{1}$$

where $Q = e^{-\frac{mhv}{kT}}$ is the activation exponential; m is the quantum subsystems number; $\Delta E_a = mhv$ is the activation energy; $D = 1 - Q^{-1}$; $A = \Omega_1^m - 1$; $B = \Omega_2^m - \Omega_1^m$; $C = \Omega_3^m - \Omega_2^m$; Ω_1, Ω_2 and Ω_3 and are the different degrees of degeneracy of the vibrational levels groups that are given by classification parameters ξ, η and ζ (Figure 2).

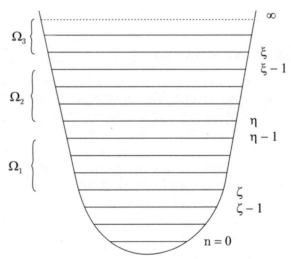

Figure 2. Classification of the vibrational levels of the oscillator taking into account of the degree of degeneracy

On the other hand, the elementary act average energy for the Boltzmann-Arrhenius model in which an activation barrier appears to be two quantum harmonic oscillators having one point of crossing (a top of the barrier) has been calculated in Ref. [2]:

$$\bar{\varepsilon}_{\rho} = hv \left(\rho + \frac{1}{e^{hv/kT} - 1} \right) \cdot e^{-\frac{\rho hv}{kT}}, \tag{2}$$

where $e^{-\frac{\rho hv}{kT}}$ is the activation exponential, and ρhv is the activation energy (Figure 3).

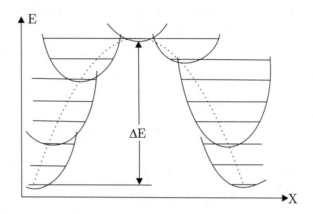

Figure 3. Activation barrier presented in the form of a transition the one quantum oscillator to another (the Boltzmann-Arrhenius model)

EXPLANATION OF IR MULTI-PHOTON ABSORPTION ON THE BASIC OF THE MODEL

To consider IR multi-photon absorption process let us re-formulate the basic theses of the model slightly: (1) during the transformation process the potential energy of a molecule changes discretely or in quanta. The transformation process appears to be a series of quantum subsystems occurring in sequence (these subsystems may also be defined as identical quantum oscillators); (2) in the intensive field of laser radiation, energy exchange between IR radiation and atoms of molecule results in discrete translation of these atoms which absorb oscillation energy by identical quanta up to molecular structure complete transformation.

Here, it is necessary to make two remarks to continue the consideration. First, we shall suppose that the number of quantum subsystems forming the activation barrier is a constant value and it does not depend on the fact, whether the molecular structure transformation occurs in equilibrium thermal radiation or in the intensive field of laser IR radiation. So the quantum subsystems number m is an internal characteristic of the molecule not depending on external conditions. Second, we shall assume that the average energy of the translationally moving atoms participating in the elementary transformation act does not depend on transformation (whether it is the equilibrium thermal transformation or the transformation owing to laser IR radiation). These two remarks lead to the following ratio:

$$\bar{\varepsilon}_m = \bar{\varepsilon}_\rho \tag{3}$$

i.e., "one big continuous jump (the Boltzmann-Arhenius model) is equal to many little discrete steps" [4]. Therefore the consideration of a molecular structure transformation in the intensive field of laser IR radiation is reduced to a arrangement of a quantum subsystems sequence m at the vibrational levels of harmonic oscillator for the Boltzmann-Arrhenius model (Figure 4) [4]. The above mentioned equality (3) is essentially a transcendental equation for the calculation of an identical quanta number ρ absorbed by the molecule under transformation of its structure in the intensive field of laser IR radiation.

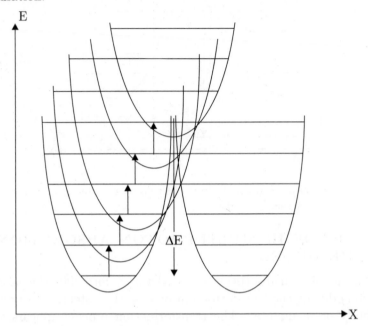

Figure 4. Coherent multi-photon absorbtion on the basis of realization of the quantum subsystems sequence

DISSOCIATION SF$_5$-F BOND

Let us carry out some quantitative estimations for the SF6 molecule on the basis of the above mentioned concepts. It is known that opticity vibration of the SF6 molecule is the vibration with the frequency $\nu_3 = 1.556\ 10^{13}$ c/s, and the dissociation energy per one chemical bond $mh\nu = 2.11$ ev [10]. Let us calculate the average energy of the translationally moving atoms participating in the elementary dissociation act at temperature $T = 298K$ $(\Omega_1 = \Omega_2 = \Omega_3 = 1)$, using the main formula (1) of the activation process model: $\bar{\varepsilon}_m = 1.361\ 10^{-37}$ ev; then, solving the transcendental equation (3) concerning the ρ parameter, one receives: $\rho = 34$, that conforms well to the experimentally observed result [11]. Using the energy value of quantum $h\nu_3 = 6.48324\ 10^{-2}$ ev, we shall define the number of quantum subsystems per dissociation of one bond S-F of the SF$_6$ molecule, taking the ratio $2.11/(6.48324\ 10^{-2}) \approx 32.55$. Therefore, approximately at half amount of the SF$_6$ molecules in ensemble the S-bond dissociates due to formation of 32 quantum subsystems; at the second half amount-dissociation occurs through 33 quantum subsystems.

Thus, the mechanism of the excitation process and subsequent dissociation of the S-F bond in the SF$_6$ molecule one can explain as the coherent multi-photon absorption on the basis of realization of the quantum subsystems sequence, in one or two of which

IR quantum is absorbed twice for formation of the subsequent quantum subsystem (Figure 4).

DISSOCIATION OF STYRENE IONS BY CO_2 LASER IRRADIATION

According to the model [3] the dissociation rate constant is defined by the expression

$$k = \frac{\bar{\varepsilon}m}{h}.$$

The numerical simulation of dissociation process for styrene ions $C_8H_8^+$ consists in the selection of ν, m, Ω_1, Ω_2, Ω_3, ξ, η and ζ parameters for Eq. (1) and comparison of temperature dependence for the constant of dissociation rate obtained and experimental results observed [12]. Next parameters have been received: $\nu = 2,818 10^{13}$ Hz (the frequency of CO_2 laser); $m = 7,5$; $\Omega_1 = 0,0527$; $\Omega_2 = \Omega_3 = 1$; $\xi = 1$; $\eta = 7$; $\zeta = 14$. To calculate the parameter ρ for the different temperatures Eq. (3) has been used (Table 1). The use of the Boltzmann-Arrhenius activation model allows to predict the dissociation rate for styrene ions with an accuracy of 30%. The accuracy of the approach used sets one thinking that there is one more mechanism of dissociation, which is probably based on somewhat different notions about the course process of elementary dissociation act. If one assumes that it results from a sequentially realized set of the metastable conformation substates, which can be approximated by the identical quantum oscillators, then it is possible to utilize the interaction model of molecule with IR-radiation. On the basis of this model the simulation of a dissociation barrier has been made by means of usage of 7,5 identical quantum oscillators with the characteristic frequency 2,818 10^{13} Hz (noninteger number of oscillators testifies that half of ions in ensemble dissociates through 7 metastable conformations substates, and other half through 8 ones). As to behaviour features of the lowest vibrational levels of oscillators (from 1 up to 6, inclusively), only 5,27% of average quantum oscillators can ensure energy exchange between themselves and surrounding IR field. For remaining 94,73% of overage quantum oscillators the transitions from the first vibrational level on the higher up to sixth are prohibited. According to this investigation when the temperature is increased the identical quanta number of laser IR radiation necessary for the elementary dissociation act of the ion grows too. Thus, the engaging of this mechanism of dissociation will allow to predict the dissociation rate of molecular ions more completely and exactly in the future by means of reasonable relation between the two mechanisms of dissociation.

Table 1. Temperature dependence for the identical quanta number of laser IR radiation necessary for the elementary dissociation act of the ion $C_8H_8^+$

$T(K)$	960	1120	1280	1440
ρ	25,56	28,62	31,695	34,784

SELF-DIFFUSION IN SILICON, GERMANIUM AND ARSENIDE OF GALLIUM

In Ref. [3] a numerical simulation has been carried out for self-diffusion processes of Si, Ge and GaAs (calculations have been made according to the formula (1)). Using the equality (3) one can calculate the identical quanta number ρ absorbed by the cluster under transformation of its structure (see Table 2). The calculation for ρ was carried out at the compensation temperatures for each of materials. It is known the compensation temperature gives a point in the Arrhenius straight line showing temperature dependence of diffusivity, that point being common for all possible experimental self-diffusion dependences in materials considered. This allows one to increase a generality degree for a forecast of solid state recrystallization processes for semiconductor layers.

Table 2. Main simulation parameters for Eq. (3)

	Si	Ge	GaAs
$\nu\ (S^{-1})$	$2,2 10^{13}$	$1,4 10^{13}$	$2,0 10^{13}$
$t_{\text{comp}}(k)$	1785	1110	1674
m	34	42	30
ρ	35	42	33
Ω_1	1	1	1
Ω_2	2	2	2
Ω_3	1	1	1
ξ	1	1	1
η	3	3	3
ζ	5	5	4

Thus, the numerical simulation carried out according to the model offered has allowed to calculate the resonant frequencies for self-diffusion processes of Si, Ge and GaAs, the quantum sub-systems number located on the activation barriers, the identical quanta number of lased IR radiation necessary for the elementary transformation of the cluster structure, the degeneration degrees and classification parameters of the quantum subsystems vibrational levels. Not only does the elementary transformation act model of cluster structure interacting with IR laser radiation allow to calculate the main parameters of the elementary activation act for non-adiabatic approximation but it also predicts a speed increase of diffusion processes. On the one hand, the elementary act can be realized due to IR quanta absorption of the certain frequency that exist in thermal equilibrium radiation. On the other hand, the elementary transformation act act course may be caused by IR laser radiation absorption. These are two incompatible events. Therefore, a resulting speed of the process course for recrystallization of ion-implanted semiconductor layers according to the equation (3) can be doubled.

PROTEIN FOLDING

It is known that low-intensity infrared laser radiation has the same action on a monomolecular chemical reaction course as black body radiation [13, 14], which is a propellent for a molecular structure transformation [15]. It would be interesting to explore this problem as applied to dynamic properties of protein mobility during its folding. Here, for numerical simulation of the first-order monomolecular chemical reaction rate constant, two models have been used: the Boltzmann-Arrhenius [2] and the activation process ones [3]. Using the conservation energy law and being aware of that an elementary activation act both in the first and in the second case requires the same amount of energy, it is possible to calculate the number of IR-radiation quantums necessary for the monomolecular chemical reaction course [4]. Earlier, on the basis of the activation process model, the kinetics of folding and insertion for the β-barrel outer membrane protein A (OmpA) of *Escherchia coli* into *dioleoylphosphatidylcholine* (DOPC) bilayers has been numerically simulated [4]. In that investigation, simulation of activation barrier has been carried out by means of eleven identical quantum oscillators with the characteristic frequency 1,056 10^{13} Hz. There, a prohibition of vibrational transitions from the ground state to the first and second vibrational energy levels has been imposed, while the third vibrational level has been fourfold degenerated. Now, applying the closed system of equations given in Ref. [4], one can solve the transcendental equation concerning a parameter describing the absorbed IR-laser radiation quantums amount. Here, this quantums amount with frequency 1,056 10613 Hz is equal to 24 at temperature of 295,5 K. As the first and second vibrational energy levels, according to the earlier carried out investigations, do not participate in energy exchange with IR-laser radiation, then its frequency should be selected triple, i.e., 3,168 10^{13} Hz. On this frequency the IR laser can exchange only with eight quantum oscillators (24 : 3 = 8). The remaining three quantum oscillators simulating intermediate metastable conformation states form from the previous quantum oscillator (or subsequent one) without absorption of the IR-laser quantum energy, that is the potential energy of the protein molecule at this stage of the conformation transformation does not alter [16]. Apparently, the quantum oscillator formation, at this stage, origins at the expense of exceeding a distance between previous and formed quantum oscillator along a generalized coordinate of chemical reaction by zero-point vibrations half-amplitude. Thus, one can double the examined reaction rate in the low-intensity IR-laser radiation.

CONCLUSION

Usually phenomena of IR multi-photon absorption and photodissociation of molecules are examined in the adiabatic approximation, which explains well resonant absorption in the first 3-4 excited vibrational levels. As to coherent multi-photon absorption in quasi-continuum and above it, the existing models of the adiabatic approximation still fail to explain the physics of these processes. In our case, the use of the activation process model allows us to make a numerical simulation in non-adiabatic approach for a significant field of application to the activation processes stimulated by IR laser radiation and receive new important information in each concrete case.

REFERENCES

1. V. L. Tavgin and A. V. Stepanov, Activation process model, Phys. Stat. Sol. (b), 161 (1990), 123–130.

2. V. L. Tavgin and A. V. Stepanov, Activation process model: application for diffusivity in covalent crystals, J. Mol. Struct. (Theochem), 257 (1992), 1–24.

3. A. V. Stepanov and V. L. Tavgin, Development of the activation process model: compensation effect, Int. J. Quantum Chem., 59 (1996), 7–14.

4. A. V. Stepanov, What can the activation process model give the adiabatic and non-adiabatic approximation? J. Mol. Struct. (Theochem), 538 (2001), 179–188.

5. A. V. Stepanov, Simulation of infrared multiphoton dissociation of styrene ions by CO_2 laser irradiation, Book of Abstracts of the 12-th International Conference on the Physics of Highly Charged Ions, Lithuania, Vilnius, 2004, A2–13.

6. A. V. Stepanov, Elementary transformation act model of cluster interacting with IR-laser radiation, Book of Abstracts of the European Materials Research Society Spring Meeting, Symposium J: Growth and Evolution of Ultrathin Films: Surface and Interface Geometric and Electronic Structure, France, Strasbourg, 2002, J/PI.05

7. A. V. Stepanov, Numerical simulation of protein folding in low-intensity infrared laser radiation, Book of Abstracts of the Eighth European Conference on Atomic and Molecular Physics, France, Rennes, 2004, 13–18.

8. J. F. Capitani, R. F. Nalewajski and R. G. Parr, Non-Born-Oppenheimer density functional theory of molecular systems, J. Chem. Phys., 76 (1982), 568–573.

9. G.Hunter, Nodeless wave functions and spiky potentials, Int. J. Quantum Chem., 19 (1981), 755–761.

10. K.S. Krasnova, Molecular Constants for Inorganic Compounds, (Russian edition), Khimiya, Leningrad, 1979, p. 24, 428.

11. V. S. Letokhov and V. M. Nauki, (Russian edition), 1 (1987), 46.

12. R.C.Dunbar, R.C.Zaniewski, Infrared multiphoton dissociation of styrene ions by low-power continuous CO_2 laser irradiation, J. Chem. Phys., 96 (1992), 5069–5075.

13. R. C. Dunbar, Kinetics of low-intensity infrared laser photodissociation, The thermal model and application of the Tolman Theorem, J. Chem. Phys., 95 (1991), 2537–2548.

14. G. T. Uechi and R. C. Dunbar, The kinetics of infrared laser photodissociation of n-butylbenzene ions at low pressure, J. Chem. Phys., 96 (1992), 8897–8905.

15. R. C. Dunbar and T. B. McMahon, Activation of unimolecular reactions by ambient blackbody radiation, Science, 279 (1998), 194–197.

16. Yu. F. Krupyanskii and V. I. Goldanskii, Dynamical properties and energy land-scape of simple globular proteins, Uspekhi Fizicheskikh Nauk (Russian), 2002, 172, 1247–1269.

Atoms and Molecules in Laser and External Fields
Editor: Man Mohan
Copyright © 2008, Narosa Publishing House, New Delhi, India

The Second-Born Approximation in Electron-Helium Collisions in the Presence of a Laser Filed

A. Makhoute and D. Khalil

UFR de Physique Atomique, Moléculaire et Optique Appliquée, Faculté des Sciences, Université Moulay Ismaïl, Meknés, Morocco

1. INTRODUCTION

Several experiments have been performed, in which the exchange of one or more photons between the electron-atom system and the laser field has been observed [1-7]. Moreover, the laser field introduces new parameters into the description of the collision such as its intensity, its frequency, and its polarization. At present, almost all the free-free experiments have been performed with a CO_2 laser as radiation field ($\hbar\omega = .117$ eV) and used Helium and Argon as atomic-target. For such cases a number of experiments have verified qualitatively the predictions of the KWA at large scattering angles[2]. In an early experiments on Argon and Helium targets, at critical geometries, where the laser polarization is almost perpendicular to the momentum transfer, Wallbank and Holmes [3-5] have however measured angular distributions larger by several orders of magnitude than those predicted by KWA. They suggested that the disagreement could be due to the polarization of the target by the field and/or its dressing effects (the effects of the internal degrees of freedom of the atom).

The aim of this work is to generalize our treatment to the case of a helium target, for which prospects of performing experiments are favorable and give new analysis about our previous works [8, 9] in particular at low collision energies where the most experiments were performed and the results are qualitatively agree with KWA. Unless otherwise stated atomic units (au) are used throughout.

2. THEORY, RESULTS AND DISCUSSION

We consider a classical monochromatic and single-mode laser field, that is spatially homogeneous, which means that it varies little over the atomic range and that the dipole approximation is valid. Working in the Coulomb gauge, we have for the vector potential of a field propagating along the $\hat{z}$ axis and represented in th1e collision plane $(\hat{x} - \hat{y})$

$$\mathbf{A}(t) = A_0 \left[\hat{x} \cos(\omega t + \varphi) + \hat{y} \sin(\omega t) \tan\left(\frac{\eta}{2}\right) \right], \tag{1}$$

with the corresponding electric field

$$\mathcal{E}(t) = \mathcal{E}_0 [\hat{x} \sin(\omega t + \varphi) - \hat{y} \cos(\omega t) \tan(\frac{\eta}{2})], \tag{2}$$

where $\mathcal{E}_0 = \omega A_0/c$, $\mathcal{E}_0$ are ω the peak electric field strength and the laser angular frequency, respectively. Here η measures the degree of ellipticity of the field and we have the particular cases of linear polarization $(\eta = 0)$ and circular polarization $\left(\eta = \dfrac{\pi}{2}\right)$ are easily recovered. Here φ denotes the initial phase of the laser field. We can recast the electric laser field in terms of its spherical components by

$$\mathcal{E}(t) = \mathcal{E}_0 \sum_{\nu=\pm 1} i\nu \hat{\mathcal{E}}_\nu \exp(-i\nu(\omega t + \varphi)), \tag{3}$$

where $\hat{\mathcal{E}}_\nu = \dfrac{1}{2} \left[\hat{x} + i\nu \hat{y} \tan\left(\dfrac{\eta}{2}\right) \right]$ is the unitary polarization vector.

In the presence of this field, we consider the elastic scattering process (electron-helium), represented by the following equation

$$e^-(k_0, E_{k_0}) + He(1^1 S) + l\hbar\omega \rightarrow He(1^1 S) + e^-(k_f, E_{k_f}), \tag{4}$$

where k_0 and k_f are respectively the momentum of the incident and scattered electrons in the presence of the laser field. $E_{k0} = k0^2/2$ and $E_{kf} = k_f^2/2$ are the projectile initial and final kinetic energies. The target helium is initially in the ground state $1^1 S$. The integer l is the number of photons transferred between the (electron-target) system and the laser field, positive values of l correspond to the absorption of photons by the system and negative ones correspond to stimulated emission photons.

The interaction between the projectile and the laser field is treated exactly and its solution is given by the non-relativistic Volkov wave function $\chi_p(r_0, t)$ [10, 11], where k is the projectile wave vector and r_0 represents the free electron coordinate. For the laser-target interaction, since we are interested by fields which have electric strengths smaller than the atomic unit $(\mathcal{E}_0 \ll 5 \times 10^9 \text{Vcm}^{-1})$ and frequencies different from the atomic transition energies, then the perturbation theory is the most appropriate method to solve the interaction process. If one restricts oneself to the first order, the 'dressed' wave function $\Phi_n(X, t)$ is well known(see [8-11]). Here $X \equiv (r_1, r_2)$ are the coordinates of the two helium target electrons and n is the principal quantum number.

Remembering that if we consider a collision kinematics, where the incident electron is fast and exchange effects are small, we shall, as a first approximation, carry out a first-Born treatment of the scattering process. The first-Born S-matrix element for the direct elastic scattering, in the presence of the laser field is given, by the expression

$$S_{el}^{B_1} = -i \int_{-\infty}^{+\infty} dt \langle \chi_{k_f}(r_0, t) \Phi_0(X, t) | V_d(r_0, X) |_{\chi_{k_0}}(r_0, t) \Phi_0(X, t) \rangle, \tag{5}$$

where $V_d(r_0, X) = -\frac{2}{r_0} + \Sigma_{j=1}^2 \frac{1}{r_{0_j}}$ is the direct electron-atom interaction potential, with $r_{0_j} = |r_0 - r_j|, \chi_{k_0}(r_0, t)$ and $\chi_{k_f}(r_0, t)$ are respectively the Volkov wave functions of the incident and scattered electrons in the presence of the laser field. $\Phi_0(X, t)$ is the 'dressed' atomic wave function describing the fundamental and final states. This type of contribution to different scattering processes has been previously computed in various instances [12-14]. By expanding the integrand in a Fourier series and integrating over t, we can recast equation (5) in the form

$$S_{el}^{B_1} = i(2\pi)^{-1} \sum_{l=-\infty} \delta(E_{k_f} - E_{k_0} - l\omega) f_{el}^{B_1,l}(\Delta), \tag{6}$$

where $\Delta = k_0 - k_f$ is the momentum transfer $f_{el}^{B_1,l}(\Delta)$, is the first Born approximation to the elastic scattering amplitude with the transfer of photons, can be split in an electronic and an atomic amplitudes. They can be written as

$$f_{el}^{B_1,\ell}(\Delta) = f_{elec}^{B_1,\ell}(\Delta) + f_{atom}^{B_1,\ell}(\Delta) \tag{7}$$

with

$$f_{elec}^{B_1,\ell}(\Delta) = -\frac{2}{\Delta^2} J_\ell(\lambda) \langle \psi_0(X) | V_d(r_0, X) | \psi_0(X) \rangle, \tag{8}$$

$$f_{atom}^{B_1,\ell}(\Delta) = f_1(\Delta) + f_2(\Delta), \tag{9}$$

$$f_1(\Delta) = -\frac{i}{\Delta^2} \sum_n \left(\frac{J_{\ell+l}(\lambda)}{\omega_{n0} + \omega} - \frac{J_{\ell-l}(\lambda)}{\omega_{n0} - \omega} \right) M_{ni} \langle \psi_0(X) | \widetilde{V}_d(\Delta, X) | \psi_n(X) \rangle \tag{10}$$

and

$$f_2(\Delta) = -\frac{i}{\Delta^2} \sum_n \left(\frac{J_{\ell+l}(\lambda)}{\omega_{fn} + \omega} - \frac{J_{\ell-l}(\lambda)}{\omega_{fn} - \omega} \right) M_{fn} \langle \psi_n(X) | \widetilde{V}_d(\Delta, X) | \psi_0(X) \rangle \tag{11}$$

where

$$\widetilde{V}_d(\Delta, X) = \sum_{j=1} 62 \exp(i\Delta.r_j) - 2 \tag{12}$$

j_ℓ is an ordinary Bessel function of order ℓ. The terms $f_{elec}^{B_1\ell}(\Delta)$ and $f_{atom}^{B_1\ell}(\Delta)$ are called, respectively 'electronic' (which correspond to the interaction of the laser field with the projectile only) and 'atomic' (which include the atomic dressing effects and thus describe the distortion of the target by the electromagnetic radiation). Here $M_{nm}^{\pm} = \mathcal{E}_0 \langle \psi_n | \hat{\mathcal{E}}_{\pm}.X | \psi_m \rangle$ are the dipole coupling matrix elements, $\omega_{nm} = E_n - E_m$ are the atomic transition frequencies, $\lambda = \Delta$. α_0 and ψ_n is a target state of energy E_n in the absence of an external field.

It should be noted that the sums over intermediate states appearing in the expressions (10) and (11) can be divided in two classes because of the selection rules arising from the matrix elements $M_{n,n'}$ Indeed, the first sum only involves intermediate states with angular momentum $\ell' = 0$ the second sum only involves intermediate states with the final angular momentum $\ell = \ell' \pm 1$, where ℓ' is the angular momentum of intermediate state.

The first-Born differential cross section, which accounts for the 'dressing' effects due to the dipole distortion of the target atom by the laser field and corresponds to the elastic scattering process, is given by

$$\left(\frac{d\sigma_{\text{el}}^{B_1,\ell}}{d\Omega}\right) = \frac{k_f}{k_0}|f_{\text{el}}^{B_1,\ell}(\Delta)|^2, \tag{13}$$

where the amplitude $f_{\text{el}}^{B_1,\ell}$ is given by equation (7).

If only the 'electronic' term retained, which ignores the 'dressing' of the target (has the familiar form obtained by studying laser-assisted potential scattering in the *first Born approximation* (FBA)), the first-Born differential cross section for elastic scattering and excitation process would be given by

$$\left(\frac{d\sigma_{\text{el}}^{B_1,\ell}}{d\Omega}\right)_{\text{no dressing}} = \frac{k_f}{k_0}|f_{\text{el}}^{B_1}(\Delta)|^2 J_\ell^2(\lambda), \tag{14}$$

where the quantity $|f_{\text{el}}^{B_1}(\Delta)|^2$ is just the field-free first-Born differential cross section corresponding to the scattering process $(0, k_0) \rightarrow (0, k_f)$.

It convenient to use the contribution of the higher order of the Born series and exchange effects for the slow incident electron, in elastic electron-atom process in the presence of a laser field. As an example, the second-order contribution to the S-matrix element for electron-helium elastic collisions from the ground state, in the direction channel and in the presence of a laser field accompanied by the transfer of ℓ photons, can be given by

$$S_{\text{el}}^{B_2} = -i \int_{-\infty}^{+\infty} dt \infty_{-\infty}^{+\infty} dt' \langle \chi_{k_f}(r_0,t)\Phi_0(X,t)|V_d(r_0,X)G_0^{(+)}(r_0,X,t;r'0,X',t')$$
$$\times V_d(r'_0 r')|\chi_{k_0}(r'0,t')\Phi_0(X',t')\rangle, \tag{15}$$

where $G_0^{(+)}$ is the causal propagator. It should be noted that this term as it stands, is second-order in the electron-atom interaction potential V_d and contains atomic wave functions corrected to first-order in the laser field strength $\mathcal{E}_0$. If one retains a global first-order correction in $\mathcal{E}_0$ for the target "dressed" states, one finds that $S_{\text{el}}^{B_2}$ is the sum of two terms which are respectively of zeroth and first-order in $\mathcal{E}_0$.

The leading matrix element, $S_{\text{el}}^{B_2 0}$, describes the collision of a Volkov electron by the undressed atom, i.e. the second-order contribution to the S-matrix element for laser-assisted collisions of zeroth-order in $\mathcal{E}_0$ is approximated in terms of a simpler second-Born amplitude by

$$S_{\text{el}}^{B_2,0} = -(2\pi)^{-1}i \sum_{\ell=-\infty}^{\ell=+\infty} \delta(E_{k_f} - E_{k_0}\ell\omega)f_{\text{el}}^{B_2,\ell,0}(\Delta), \tag{16}$$

with

$$f_{\text{el}}^{B_2,\ell,0}(\Delta) = J_\ell(\lambda)f_{\text{el}}^{B_2}(\Delta), \tag{17}$$

where

$$f_{\text{el}}^{B_2}(\Delta) = -\frac{1}{\pi^2} \int_0^{+\infty} q^2 dq d\xi_q' \frac{\langle \psi_0 | \tilde{V}_d(\Delta_f, X) G_c(\xi') \tilde{V}_d(\Delta_0, X) | \psi_0 \rangle}{\Delta_0^2 \Delta_f^2} \tag{18}$$

is the filed-free second-Born elastic amplitude evaluated at the shifted momenta Δ_0 and Δ_f. Here $G_c(\xi') = \sum_n \frac{|\psi_n\rangle\langle\psi_n|}{\xi'-E_n}$ is the Coulomb Green's function with argument $\xi' = E_{k_i} + E_0^{He} - E_q + \ell\omega$, where $E_0^{He} = -2.904$ au is the ground state energy of the helium target and E_q is the virtual projectile energy.

In the same way, the contribution to the S-matrix element for laser-assisted collisions of first-order in $\mathcal{E}_0$, is given by, by shifting the pole of the integrand, respectively, below the real ω-axis by a small positive quantity $\mathcal{E} \to 0^+$,

$$S_{\text{el}}^{B2,1} = -(2\pi)^{-1} i \sum_{l=-\infty}^{\infty} \delta(E_{k_f} - E_{k_0} - \ell\omega) f_{\text{el}}^{B_2,l,1}(\Delta), \tag{19}$$

with

$$f_{\text{el}}^{B_2,\ell,1}(\Delta) = i J_\ell'(\lambda)[f_1(\Delta) + f_2(\Delta) + f_3(\Delta)], \tag{20}$$

where

$$f_1(\Delta) = -\frac{1}{(2\pi)^2} \sum_{n,n'} \int dq \frac{f_{0,n'}^{B_1}(\Delta_f) f_{n',n}^{B_1}(\Delta_0) M_{n,0}}{(E_q - E_{k_0} + \omega_{n',0} - i\mathcal{E})\omega_{n,0}}, \tag{21}$$

$$f_2(\Delta) = -\frac{1}{(2\pi)^2} \sum_{n,n'} \int dq \frac{M_{0,n'} f_{n',n}^{B_1}(\Delta_f) f_{n,i}^{B_1}(\Delta_0)}{\omega_{n',0}(E_q - E_{k_0} + \omega_{n,0} - i\mathcal{E})} \tag{22}$$

and

$$f_3(\Delta) = -\frac{1}{(2\pi)^2} \sum_{n,n'} \int dq \frac{f_{0,n'}^{B_1}(\Delta_f) M_{n',n} f_{n,0}^{B_1}(\Delta_0)}{(E_q - E_{k_0} + \omega_{n',0} - i\mathcal{E})(E_q - E_{k_0} + \omega_{n,0} - i\mathcal{E})}. \tag{23}$$

The study of second-order corrections to atomic $s - p$ amplitudes show that these corrections tend to a constant value of order k_0^{-1} as Δ becomes small, i.e. at small scattering angle and thus are rather unimportant in this angular range. However, this is precisely the scattering angular region, which we interest for $\mathcal{E}_0 \ll 1$ au, because the first-order amplitude is adequate to provide a significant 'dressing' effects, which supply a contribution of order Δ^{-1} and thus rule the differential cross section, while at larger scattering angles the target 'dressing' becomes less important, and under non-resonant conditions one also can model the atom by a structureless center of force. For this, in our previous works, we have neglected the second-order contribution to the S-matrix element for laser-assisted collisions calculated in first-order in $\mathcal{E}_0$ (for more detailed analysis, see Refs. [8-9]. When this approximation is adopted, we may concentrate our discussion on the computation of the dominant term, $S_{f,i}^{B_2,0}$ in laser-assisted collisions and its computation. Thus, the electron-atom interaction amplitudes with the transfer of ℓ photons may be written, in SBA, as

$$f_{\text{el}}^{\ell}(\Delta) = f_{\text{el}}^{B_1,\ell}(\Delta) + f_{\text{el}}^{B_2,l,0}(\Delta), \tag{24}$$

where $f_{\text{el}}^{B_1,\ell}(\Delta)$ and $f_{\text{el}}^{B_2,\ell,0}(\Delta)$ are respectively, the first-order and second-order amplitudes, which are given by equations (7) and (17). In other case, i.e., when the scattering angle is not small, the S-matrix element contribution for laser-assisted collisions of first-order in $\mathcal{E}_0$, becomes significant and the equation (24) can be written in the form

$$f_{\text{el}}^{\ell}(\Delta) = f_{\Delta}^{B_1,\ell}(\Delta) + f_{\Delta}^{B_2,\ell,0}(\Delta) + f_{\Delta}^{B_2,\ell,1}(\Delta), \tag{25}$$

where $f_{\text{el}}^{B_2,\ell,1}(\Delta)$ represents the first-order correction in $\mathcal{E}_0$ of the second-order scattering amplitude, which is given by expression (20)

In the calculation of the amplitudes (7), (17) and (20), we need to know the explicit form of the atomic wave functions in the absence of an external field. For the ground state of helium and for n^1S, n^1P, n^1D we use the wave functions proposed in Ref.[15]. We note that the doubly excited states are neglected in view of the weak contribution of these states to the elastic process [16].

The main problem in evaluating the scattering amplitudes corresponding to the second-order contributions to the S-matrix element for laser-assisted elastic scattering and excitation process (see Refs. [8] and [9])

The contribution for laser-assisted elastic collisions to the S-matrix of exchange scattering which leads to some conceptual diculties but would not significantly alter the results of the present discussion. We have consider in the present paper only the landing term of $g_{\text{el}}^{B_1,\ell}$, the exchange amplitude for electron-atom collisions with the transfer of ℓ photons used in Ref. [16]. It is known that the exchange effects in collisions are important at low relative velocities, while the FBA is an essentially high-energy approximation. Thus, the FBA does not seem the best approximation to look into the effects we are interested in. In favor of the FBA there are, however, the relative simplicity of the analytical treatment.

The second-Born differential cross section corresponding to the elastic scattering process, with the transfer of ℓ photons, is given by

$$\frac{d\sigma_{\text{el}}^{\ell}}{d\Omega} = \frac{k_f}{k_0}\left[\frac{1}{4}|f_{\text{el}}^{\ell} + g_{\text{el}}^{B_1,\ell}|^2 + \frac{3}{4}|f_{\text{el}}^{l} - g_{\text{el}}^{B_1,\ell}|^2\right]. \tag{26}$$

does not depend on the initial phase φ of the laser field, due to the inability of the collision time to be defined, as a result of the approximation of the projectile wave packet by a mono energetic beam of infinite duration [17].

When comparing theoretical results to the experimental cross sections, one first has to obtain cross sections over a fine mesh of intensities and then convolute them with a realistic spatiotemporal distribution of intensities of the laser beam. The ponderomotive acceleration of the projectile when penetrating and leaving the interaction region should also be taken into account, unless the pulse is extremely short.

The present semiperturbative method with the sturmain basis expansion takes into account the target atom distortion induced by the presence of a laser field. The validity of our treatment is based on the fact that the laser-helium target interaction is non resonant. The condition on the interesting is more stringent if the laser frequency is comparable to any characteristic atom transition frequency. We note that the elastic scattering process can be considered as non resonant if for a given frequency, the intensity does not exceed a certain limit[15]. Such a condition will be respected by our choice of the Nad-YAG laser frequency $\omega = 1.17$ eV and $\mathcal{E}_0 = 2. \times 10^7$ V cm^{-1}.

We are interested in demonstrating the effects of the incident electron energies in the elastic collision of fast electrons by a helium target in the presence of a laser field. In helium, there have been comparatively fewer attempts to address the role of the dressing of the atomic states by the strong laser field, which not completely answered yet, the main reason being that the computation is much more complex. This unfortunate in view of the fact that helium would lend itself more easily than hydrogen, to experimental verification. Note however that, through simplified, the model contains all the ingredients needed for the discussion of the physics of such processes. Our results are interpreted by reckoning the first and second Born differential cross sections, where the elastic field strength and laser photon energy are kept fixed. We have checked as a function of scattering angles our treatment in SBA and FBA, and Kroll-Watson approximation give similar results at 100 eV of the incoming electron energy. This also happens for no forward scattering angles and any collision energy when we compare the FBA and KWA. Thus it proves that a higher-order of Born series character of the laser-atom interaction is important for a proper analysis of any scattering angles free-free transitions in a low incident electron energies, and that the main reason for the discrepancies between the experimental data and/or the Kroll-Watson approximation is dressing effects character, in particular, of the forward scattering angle.

The results presented in this paper are obtained for a geometry in which the polarization vector of the field $\mathcal{E}_0$ is parallel to the direction of incoming electron wave vector k_i, where free-free differential cross sections are maximum at a particular laser intensity and incident electron energy[6] and where the laser-assisted differential cross section only depends on the orientation of the polarization unit vector $\hat{\mathcal{E}}_\nu$ [18]. We compare our results in SBA with the obtained in FBA and with the values obtained by using the *Kroll-Watson Approximation* (KWA), where the differential cross sections for exchange of photons are related to the field-free differential cross section $\left(\frac{d\sigma}{d\Omega}\right)$ through

$$\frac{d\sigma^\ell}{d\Omega} = \frac{k_f}{k_i} J_\ell^2(\lambda) \frac{d\sigma}{d\Omega} \tag{27}$$

In set of Figures 1-3, we present differential cross sections for laser-assisted elastic scattering with net exchange of up to two photons ($\ell = 0, \pm 1$) when the electron is incident along the laser polarization axis at collision energies of 4.5eV, 10eV and 20eV. As is already noted by several authors [17–20], the dressing effects are seen to be dominant in the forward direction. This is due to the presence in the 'atomic' term in FBA of $s - p$ transition amplitude with behavior like K^{-1} for small transfer momentum K. Moreover, we notice a destructive interference, in FBA, between the 'electronic' and 'atomic' amplitudes. The presence of such interference is a general feature of $1s \rightarrow np$ transitions in the case of inverse bremsstrahlung ($\ell > 0$). This minimum appears at angles for which the first Born differential cross vanishes, i.e. when $f_{\text{elec}}^{B_1,\ell}(\Delta) + f_{\text{atom}}^{B_1,\ell}(\Delta) = 0$. In contrast, for the cases when the zeroth-order term of the SBA to the elastic scattering amplitude is taken into account,we do not obtain the deep minimum, because the presence of an additional term forbids the occurrence of a complete destructive interference.

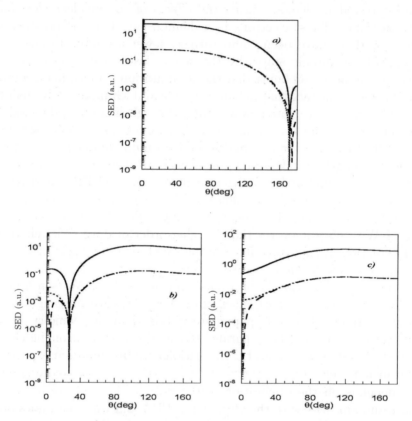

Figure 1. Differential cross section for elastic scattering with the transfer of ℓ photons [(a) $\ell = 0$, (b) $\ell = 1$, (c) and $\ell = -1$, (d)] as a function of scattering angle (θ). The incident electron energy is 4.5 eV, the laser frequency is 1.17 eV and the electric field strength is 2×10^7 V cm^{-1}. Solid lines: Second-Born approximation. Dashed lines: First-Born approximation. Dotted lines: results obtained by neglecting the dressing of the target

In Figures 1-3, we show in more detail, the variation of the SBA differential cross section for a given intensity and for various photon number corresponding to the absorption and emission. The second-Born approximation results show a behavior completely different for the differential cross sections, as compared with those obtained from the results of calculations in first-Born approximation and in which dressing effects are neglected. This behavior of the cross section in SBA is similar to that provided by measurements of the differential cross section for scattering of electrons, from helium, in the presence of a high-intensity CO_2 laser field, such that the new relative minima appears at different scattering angles, as in the case of without net exchange of photons. These minima become less marked and lower when the energy of the incident electron increases. One observes, for a given laser intensity and frequency, the disappearance of the minima, corresponding to values of the scattering angle, when the incoming electron energy less than $\simeq 10$ eV for which the cross section is actually zero. For the net exchange of one photon, these minima appearing in FBA cross section and resulting from a destructive interference between 'electronic' and 'atomic' terms exist for any incident energies which are acceptable physically. These minima becomes shallower as the collision energy increases, so that the overall difference with the emission cross section becomes smaller. For the net exchange of one photon, the emission cross section are slightly larger than those for absorption, while for exchange of two photons the absorption cross sections are slightly larger than those for emission. Moreover, the one-photon differential cross sections are larger than those for the two-photon processes; for in excess of the values calculated by FBA and KWA near predicted by the experimental data [6].

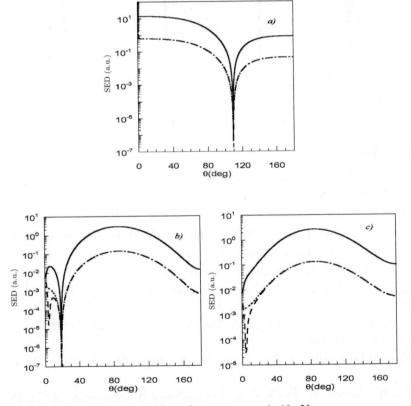

Figure 2. As Figure 1, but with the incident electron energy is 10 eV

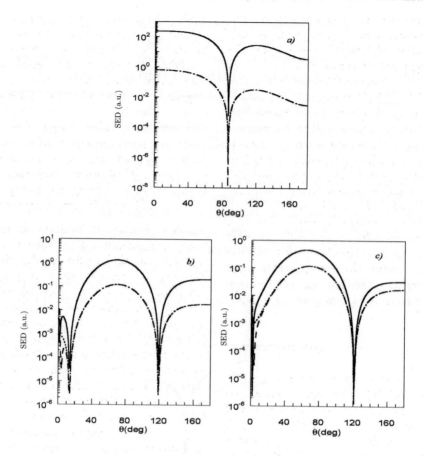

Figure 3. As Figure 1, but with the incident electron energy is 20 eV

REFERENCES

1. M. N. Kroll and K. M. Watson (1973), Phys. Rev. A, 8, 804.

2. N. J. Mason (1993), Rep. Prog. Phys., 56, 1275.

3. B. Wallbank and J. K. Holmes (1993), Phys. Rev. A, 48, R2515.

4. B. Wallbank and J. K. Holmes (1994), J. Phys. B: At Mol. Opt. Phys., 27, 1221.

5. B. Wallbank and J. K. Holmes (1994), J. Phys. B: At. Mol. Opt. Phys., 27, 5405.

6. B. Wallbank and J. K. Holmes (2001), Can. J. Phys., 79, 1237.

7. B. Wallbank, J. K. Holmes and A. Weingartshofer (1987), J. Phys. B: At. Mol. Opt. Phys., 20, 6121.

8. M. Bouzidi, A. Makhoute, D. Khalil, A. Maquet and C. J. Joachain (2001), J. Phys. B: At. Mol. Opt. Phys., 34, 737.

9. A. Makhoute, D. Khalil, M. Zitane and M. Bouzidi (2002), J. Phys. B: At. Mol. Opt. Phys., 35, 957.

10. D. Khalil, A. Maquet, R. Taïeb, C. J. Joachain and A. Makhoute (1997), Phys. Rev. A, 56, 4918.

11. A. Makhoute, D. Khalil, A. Maquet, C. J. Joachain and R. Taïeb (1999), J. Phys. B: At. Mol. Opt. Phys., 32, 3255.

12. A. Maquet and J. Cooper (1987), Phys. Rev. A, 41, 1724.

13. M. Bouzidi, A. Makhoute and M. N. Hounkounou (1999), Eur. Phys. J. D, 5, 159.

14. O. Khalil El Akramine, A. Makhoute, A. Maquet and R. Taïeb (1998), J. Phys. B: At. Mol. Opt. Phys., 31, 1115.

15. P. Francken, Y. Attaourti and C. J. Joachain (1988), Phys. Rev. A, 38, 1785.

16. P. Francken and C. J. Joachain (1987), Phys. Rev. A, 35, 1590.

17. M. Dorr, C. J. Joachain, R. M. Potvliege and S. Vucic (1994), Phys. Rev. A, 49, 4852.

18. F. W. Byron, Jr., P. Francken and C. J. Joachain (1987), J. Phys. B: At. Mol. Opt. Phys., 205487.

19. D. Khalil El Akramine, A. Makhoute, A. Maquet and R. Taïeb (1998), J. Phys. B: At. Mol. Opt. Phys., 31, 1115.

20. O. El Akramine, A. Makhoute, D. Khalil, A. Maquet and R. Taïeb (1999), J. Phys. B: At. Mol. Opt. Phys., 32, 2783.

21. S. Geltman (1995), Phys. Rev. A, 51, R34.

22. I. Rabadán, L. Méndez and A. S. Dickinson (1994), J. Phys. B: At. Mol. Opt. Phys., 27, L535.

Atoms and Molecules in Laser and External Fields
Editor: Man Mohan
Copyright © 2008, Narosa Publishing House, New Delhi, India

Observation of DIP in Rate of Ionization in a Discharge under Externally Applied Magnetic Field

R. Bordoloi[1] and G. D. Baruah[2]

[1] Department of Physics, Dibrugarh University, Dibrugarh 786 004, India
[2] Department of Physics, Tinsukia College, Tinsukia 786 125, India

1. INTRODUCTION

In a steady discharge the discharge current is mainly governed by the secondary ions produced in the cavity and in the electrodes of the tube [1]. Among various factors contributing to the secondary ionization, the electron impact is by far the most significant one. Production of charge carriers by secondary processes in a discharge tube are more predominant and are of more importance than the primary processes. However it should also be realised that the ions produced in the primary processes are solely responsible for secondary ionisation. Among many possible causes of secondary ionisation some of the most prominent factors are: primary electrons, positive ions, excited or metastable atoms (We shall use the word atom to represent both atoms and molecules.) and photons etc.

Townsend [2] in his study on secondary ionisation showed that when a gap of length d between two parallel electrodes in air was irradiated, the current increased due to multiplication of the ions between the plates. Later work on this showed that the negative ions were effective in multiplication, and that when the cathode only is irradiated, the current i flowing in the gap is given by

$$i = i_0^e \alpha (d - x_0) \tag{1}$$

Where i_0 is the photoelectric current from the cathode and α is the first Townsend ionisation coefficient (It gives the number of ion-pairs produced per cm. in the direction of the electric field of strength X. If p is number of ion-pairs produced per second in the volume between the plates, then we see that α/p is a function of x/p for a given gas, and x_0 is the distance travelled before the electrons acquire sufficient energy to ionise. It is also found that when the gap d between the electrodes is made very large the graphs of $\log(\frac{i}{i_0})$ against d curve up defying the prediction of the Eqn. (1). This happens so because in this situation the positive ions already produced in the tube become another source of ionisation, as they get enough free-path to gain sufficient kinetic energy. In order to account for the ionisation by the positive ions Townsend introduced a new coefficient β, which represents the number of ion-pairs formed by a positive ion in moving 1 cm in field direction. In terms of the new coefficient β the modified form of Equ. (1) becomes:

$$\frac{i}{i_0} = \frac{(\alpha - \beta)e^{(\alpha-\beta)d}}{\alpha - \beta e^{(\alpha-\beta)d}} \tag{2}$$

And for $\beta \ll \alpha$ it becomes

$$\frac{i}{i_0} = \frac{(\alpha - \beta)e^{(\alpha-\beta)d}}{1 - (\beta/\alpha)e^{(\alpha-\beta)d}} \tag{3}$$

However modification of the above equations for $\alpha \sim \mu$, $\alpha \ll \mu$ and $\alpha \gg \mu$ as was done by Townsend led us to the conclusion that the equations representing the volume ionization of the gas can't account for the up curving of the $\log(\frac{i}{i_0})$ vs. d curve. This shows that photo ionisation in the cavity can explain the experimental results only when the absorption coefficients μ and μ_i neither too large nor too small (because then the absorption will be too small to cause any ionisation). Such a huge constraint makes it extremely difficult for the volume ionization [4] to give a major contribution to the discharge current. However the result obtained in our work points exactly to a different situation in which the ionization in the cavity of the tube seems to have a major role in the discharge current.

2. EXPERIMENT

The experimental arrangement is as follows. In a conventional π-type discharge tube made of hollow Aluminium electrodes, a very low pressure is maintained with the help of continuously operating vacuum pump. Electrode voltage operating just above threshold ensures a continuous discharge with a bluish white colour representing the Angstrom band of the CO. The discharge current is being measured with a sensitive digital meter and is found to be constant. Two different discharge tubes were used. Tube No. 1 has a cavity length of 30 cm. and tube No. 2 has a cavity length of 20 cm.

Now the cavity of the tube is subjected to a strong transverse magnetic field of the order of Kilo Gauss. The discharge splits into two distinct paths. The new discharge current is then measured with the same digital meter.

3. CONCLUSION

The result of the experiment bears a great deal of significance. As is observed from the table.1, the discharge current, on application of the magnetic field decreases to almost 37%of the initial value in tube No. 1 and to 80% of its normal value in tube No. 2.

Table 1. Change in discharge current by magnetic field

S1 No.	Discharge current without magnetic field	Time in Sec.	Discharge current in presence of the magnetic field	Remark
1	Tube No. 1	1 Sec	6.2	Initially the current goes on decreas-
2	(30 cm length)	2 Sec	5.8	ing with times, in a very quick time
3	7.5 mA	3 Sec	3.7	it becomes constant. However after
4		3.5 Sec	2.8	the field is removed, the current again
5		4 Sec	2.8	rises to the original value.
6		5 Sec	38	
1	Tube No. 2	1 Sec	3.0	Decrease in current is smaller than
2	(20 cm length)	2 Sec	2.8	the other tube. After the field is
3	3.2 mA	3 Sec	2.7	removed, the current again rises to
4		3.5 Sec	2.5	the original value.
5		4 Sec	2.5	
6		5 Sec	2.5	

At the same time the discharge splits into two distinct paths at the centre of the cavity in each case. Also it has been noticed that the dip in current occurs only when the discharge splits i.e. only when the magnetic field is applied transversely. In fact the decrease in discharge current and the splitting of the discharge are interrelated. As we know that the inelastic collision between the highly accelerated electrons and the positive ions or the metastable atoms produce secondary ions [4], [5] in the cavity. The number of secondary ions produced in the cavity depends upon X/p, the ratio between the electric field and the gas pressure and the length of the cavity. Due to splitting of the discharge, the ions of opposite kinds follow separate path through the cavity. This prevents them from producing secondary ions in the cavity as collision of the electrons with the ions or the excited atoms is no more possible. Hence the current drops down. It has also been observed that in the second tube though all the operating conditions were identical, the discharge current was relatively lower than in the first tube and also the drop in current due to the application of the magnetic field was smaller. This was due to shorter length of the cavity in the second tube. As the shorter length reduced the mean free path of the molecules which in turn reduced the probability of inelastic collision and that finally led to decrease in on production. In fact from the results of our work it can be seen that the influence of the "cavity ionization" on discharge current is more pronounced in a longer tube than in a shorter one.

REFERENCES

1. P. F. Little, Encyclopaedia of Physics, Vol. XXI, 574.
2. Encyclopaedia of Physics, Vol. XXII.
3. J. Dutton *et. al.*, Proc. Roy. Soc. London Sec. A, 218, 206 (1953).
4. A. von Engel, Encyclopaedia of Physics, Vol. XXI.
5. H. S. W. Massey, Encyclopaedia of Physics, Vol. XXXVI.

Atoms and Molecules in Laser and External Fields

Editor: Man Mohan

TL Study of $Ca_xSr_{1-x}SO_4$: Eu Phosphors

S. P. Lochab[1], P. D. Sahare[2], Numan Salah[2], R.S. Chauhan[3], Ranju Ranjan[2], A. Pandey[4] and Amitanshu Patnaik[5]

[1] Inter University Accelerator Center, Aruna Asaf Ali Marg, New Delhi 110 067, India
[2] Department of Physics and Astrophysics, University of Delhi, Delhi 110 007, India
[3] Department of Physics, RBS College, Agra 282 002, India
[4] Department of Physics, Sri Venkateswara College, University of Delhi, New Delhi 110 002, India
[5] Laser Science and Technology Centre, Metcalfe House, DRDO, Delhi 110 054, India

1. INTRODUCTION

In radiation dosimetry, sulphates have been popular phosphors for *thermoluminescence* (TL) [1-8]. Among them calcium sulphate is more sensitive TL phosphor. In the family of calcium sulphate phosphors $CaSO_4$: Dy is most sensitive. $SrSO_4$ has also been a favourable phosphor in this field. Several works have been done on sulphates of calcium and strontium doped with different rare earth materials.

Of late mixed sulphates have created interest among the workers in the field of dosimetry and a lot of work has been done on them [9-12]. In the present paper we are interested in TL of a mixed sulphate $Ca_xSr_{1-x}SO_4$: Eu with x varying in the range between 0 to 1 (for $x = 0, 1$ the study has yet to be done).

2. EXPERIMENTAL

$Ca_xSr_{1-x}SO_4$: Eu samples were prepared by acid method taking AR grade calcium sulphate and strontium sulphate in different ratios with 0.2mol% AR grade Europium oxide. On distilling the acid away the phosphor was obtained and was washed several times by double distilled water and then dried. The dried sample was ground to 100 to 200 μm grain size and kept for annealing in quartz crucible at 973 K for 1hr in air.

Samples of this phosphor were exposed to γ-rays of a Cs^{137} source for various exposures (0.10 Gy to 100 Gy) at room temperature. TL glow curves were recorded on a Harshaw TLD reader model 3500 by taking 5mg of the samples each time. Also the glow curves of $CaSO_4$: Dy powders were recorded for comparison of the sensitivity and the structure under identical conditions.

3. RESULTS AND DISCUSSION

3.1. TL Response

Figure 1 shows the TL glow curves of $Ca_xSr_{1-x}SO_4$: Eu with x having values 0.1, 0.3, 0.5 and 0.7. The TL sensitivities of these four compounds were compared with the standard phosphors $CaSO_4$: Dy. It is clear from the figure that our compound has shown its highest sensitivity at $x = 0.1$.

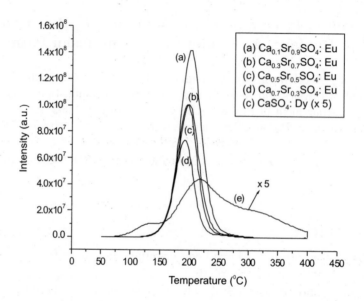

Figure 1. TL glow curves of various phosphors exposed to 10 Gy of gamma-ray

A change in the crystal structure of the phosphors $Ca_xSr_{1-x}SO_4$:Eu on varying x is possibly the reason for this variation in TL sensitivity and the crystal structure of the phosphors for $x = 0.1$ is presumably such that it favors formation of more defects which leads to an enhanced TL sensitivity. Also it has been seen that Eu present in its Eu^{2+} state leads to a higher sensitivity than that in its Eu^{3+} state [13, 14]. Thus it is quite possible that in the phosphors $Ca_xSr_{1-x}SO_4$:Eu the dopant is predominantly present as Eu^{2+}, resulting in higher TL emission. A PL study for these compounds is being done to prove this speculation.

3.2. Other Dosimetric Characteristics

The TL response of the phosphor $Ca_{0.1}Sr_{0.9}SO_4$: Eu has been shown in the Figure 2. It is found to be linear in the dose range 0.01 to 50 Gy. Beyond this range the response becomes sublinear. The phosphor is found to show excellent reusability as can be seen in Figure 3. Fading has also been found to be quite less. A complete analysis can be made from Figure 4. In the dose range 0.10 to 100 Gy, the glow curve shape and structure also do not change.

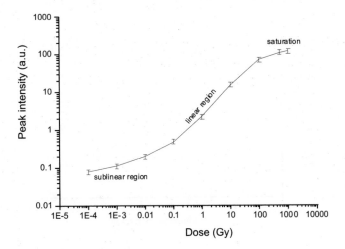

Figure 2. TL response of $Ca_{0.1}Sr_{0.9}SO_4$: Eu

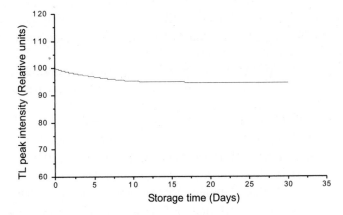

Figure 3. Fading in the peak intensity over a period of time

4. CONCLUSION

It has been shown that for $x = 0.1$ the phosphor $Ca_x Sr_{1-x} SO_4$: Eu shows maximum TL sensitivity. Moreover, the phosphor $Ca_{0.1}Sr_{0.9}SO_4$: Eu has also been found to have negligible fading and excellent reusability with simple glow curve structure which does not change with doses. It shows linear response to a wide range of gamma dose, making it suitable for TL dosimetry.

Acknowledgements

We are grateful to Prof. Prateek Kumar, A.I.I.M.S., New Delhi for helping us using the facilities at A.I.I.M.S. We are also highly thankful to Dr. Amit Roy, Director, I.U.A.C., New Delhi, for the encouragement and support that he gave to us.

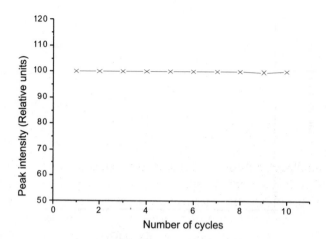

Figure 4. Reusability of $Ca_{0.1}Sr_{0.9}SO_4$: Eu

REFERENCES

1. E. Wiedemann and G. C. Schmidt, Ann. Phys. Chem., 54 (1895), 604.
2. T. Yamashita, N. Nada, H. Onishi and S. Kitamura, Proc. 2nd Internat. Conf. Lumin. Dosimetry, Gatlinburg (USA), 1968, p. 4.
3. T. Yamashita, N. Nada, H. Onishi and S. Kitamura, Health Phys., 21 (1971), 295.
4. T. Yamashita, Proc. 4th Internat. Conf. Lumin. Dosimetry, Krakow (Poland), 1974, p. 467.
5. V. N. Bapat, J. Phys. C: Solid State Phys., 10 (1977), L465-L467.
6. S. V. Godbole, J. S. Nagpal and A. G. Page, Radiat. Meas., 32 (2000), 343.
7. R. L. Dixon and K. E. Ekstrand, Phys. Med. Biol., 19 (1974), 196.
8. U. Madhusoodhan, M. T. Jose and A. R. Lakshmanan, Radiat. Meas., 30 (1999), 65.
9. P. D. Sahare and S. V. Moharil, Radiat. Eff. Defects Solids, 114 (1990), 167.
10. P. D. Sahare and S. V. Moharil, Radiat. Eff. Defects Solids, 116 (1990), 275.
11. P. D. Sahare, S. V. Moharil and J. Lumin., 43 (1989), 369.
12. P. D. Sahare, S. V. Moharil and B. D. Bhasin, J. Phys. D: Appl. Phys., 22 (1989), 971.
13. S. V. Upadeo and S. V. Moharil, J. Phys. D Appl. Phys., 7 (1995), 957.
14. A. Pandey, R. G. Sonkawade and P. D. Sahare, J. Phys. D: Appl. Phys., 35 (2002), 2744.

Atoms and Molecules in Laser and External Fields

Editor: Man Mohan

Copyright © 2008, Narosa Publishing House, New Delhi, India

Laser Produced Spectrum of Zn₂ Molecule in the 380-450 nm Region

Subhash C. Singh, K. S. Ojha and R. Gopal

Laser and Spectroscopy Laboratory, Department of Physics, University of Allahabad, Allahabad 211 002, India

1. INTRODUCTION

In recent years spectroscopic studies both theoretical and experimental of Vander Waals and excimer molecules such as Hg_2, Cd_2 and Zn_2 are of continuing fundamental interest and also relevant to the possible development of new high power excimer laser system. Both the rare gas and group II B metal dimers have repulsive ground state with only shallow Vander-Waals minima. Lasing is thus possible in a "bound-continuum" transition with the dissociative ground state insures the condition of population inversion. Recently Laser Produced Spectroscopy has provided an advance technique for recording the spectra. The first spectra of zinc molecule were reported by Winans [1] in absorption in 1929 while by Winans [2], and Hamada [3] in 1931 in emission. Hamada produced Zn_2 in a hollow cathode discharge and reported only three band maxima lying at 368.0nm, 378.8nm and 475nm. Ault et al [4] have reported UV absorption spectra of $Zn + 2$ molecule in Ar and Kr matrices and assigned 252.0nm band to $^1\Sigma_u^+ \leftarrow X^1\Sigma_g^+$ transition. Czajkowski et al [5] have reported first excitation spectrum of Zn_2 in supersonic expansion beam crossed with a laser beam that gives information about the $X0_g^+(^1\Sigma_g^+)$ ground state and $0_u^+(^3\prod_u)$ excited state. Hay et al [6], Bender et al [7] and Couty et al [8] have calculated potential energy curves for diatomic zinc molecule and suggested $^1\Sigma_g^+$ as a repulsive ground state of zinc dimer, while most of the excited states are attractive and few are repulsive. Recently Kedzierski et al [9-15] have recorded six band systems of Zn_2 molecule in UV region using LIF technique. Kedzierski et al have also recorded excitation spectra of Zn_2 molecule in the 628-720nm [9], 530-570nm[10] and 420-470 nm [11], out of which last one has been assigned as $0_u^+(^1\Sigma_u^+) \leftarrow \prod_g$ transition. In the present paper studies of Zn^2 spectra recorded by laser vaporization of zinc in 380-455 nm region has been reported.

2. EXPERIMENTAL SETUP

The Experimental system consist of a pulsed Nd-YAG (Spectra Physics, USA) laser with 10 Hz, an ablation chamber, and a computer controlled Spex TRIAX 320 m monochromator fitted with TE cooled ICCD detector system. The zinc rod (Spec- pure Johnson Mathey, UK) was placed inside a laser ablation chamber, which is mounted on the throat of rotary and diffusion pumps. The ablation chamber was evacuated up to 10^{-3} torr using rotary pump and filled with argon gas at one torr pressure. The zinc plasma was produced using 532 nm of Nd: YAG laser with energy of 35 mJ. The zinc target was continuously rotating and translating so that each laser pulse fell on a fresh surface. The produced plasma was allowed to cool adiabatically for about 200 ns by giving delay time for ICCD gating. Radiation from the cooled plasma was focused on the entrance slit of computer-controlled monochromator fitted with TE cooled ICCD using a cylindrical lens of focal length 25 cm. Signals from the ICCD were sent to the computer using Spectra-Max Software. The Galactic Grams 32 software was used for data acquisition and spectra analysis. The experimental set-up used for recording spectra has been shown in Figure 1.

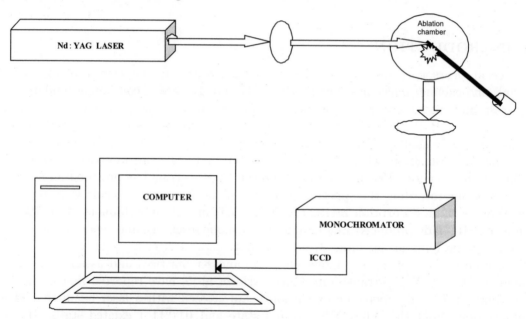

Figure 1. Experimental setup

3. RESULTS AND DISCUSSION

Laser Produced Spectra of Zn_2 molecule recorded in the region 380-455nm consists of 74 bands out of which 63 are new while rest are reported by earlier workers. Recorded spectra have been classified into two new and one known system. Detailed of these band systems are given below.

3.1.   Spectrum of $E(^1\Sigma_u^+) \rightarrow A(^3\prod_g)$ Transition

This system contains twenty-two red degraded bands, which contains eleven bands reported by Kedzereski et al [11] in the region 420-470nm. The band system has been assigned to the transition $E(^1\Sigma_u^+) \rightarrow A(^3\prod_g)$. The molecular constants have been modified are very close to those reported by Kedzereski et al [11]. Vibrational constants of the system are given below.

$$\nu_{00} = 23141.6 \text{cm}^{-1}, \quad \omega_e' = 185.0 \quad \omega_e'X_e' = 1.8 \quad \omega_e'' = 222.0 \quad \omega_e''X_e'' = 1.5$$

Spectrum of the system has been displayed in Figure 2.

3.2.   Spectrum of $J(^3\Sigma_g^+) \rightarrow D(^1\prod_u)$ Transition

This is a new system lying in the region 430-455 nm consisting of 20 violet degraded bands. These bands have been attributed to $\Delta v = 0, \pm 1, \pm 2, \pm 3$ and $\pm$ 4 sequences with ν_{00} lying at 443.4nm. Hay [6], and Couty [7] have shown that P.E. curve for $D(^1\prod_u)$ arise from $(4^1P + 4^1P)$ lies at 36619.64 cm^{-1} while the $J(^3\Sigma_g^+)$ arise from $(4^3P + 4^3P)$ lies at 58999.6 cm^{-1}. The difference of these potential energies comes out to be 22379.96 cm^{-1} which is very close to the ν_{00} in the present system. This shows that present band system arises from the transition $J(^3\Sigma_g^+) \rightarrow D(^1\prod_u)$. The vibrational constants for $D(^1\prod_u)$ and $J(^3\Sigma_g^+)$ determined in the present case are very close to those observed by Kedzerski et al in the case of $D \rightarrow A$ [9] and $J \rightarrow B(^3\Sigma_g^+)$ [15] systems. Molecular constants determined are as follows.

$$\nu_{00} = 23141.6 \text{cm}^{-1}, \quad \omega_e' = 161.0 \quad \omega_e'X_e' = 0.7 \quad \omega_e'' = 146.7 \quad \omega_e''X_e'' = 1.90$$

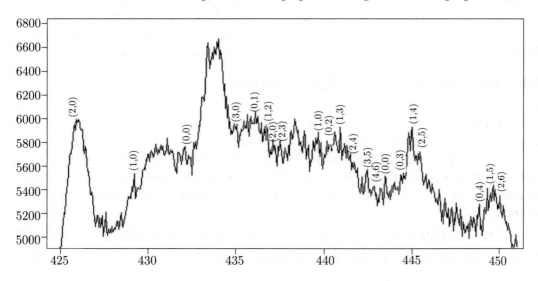

Figure 2. Laser produced spectrum of $E \rightarrow A$ transition

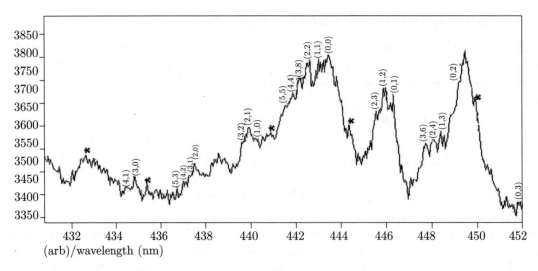

Figure 3. Laser produced spectrum of $J \to D$ transition

3.3.   Spectrum of $F \to A(^3\prod_g)$ Transition

Thirty-two red degraded bands lying in the region 380-420nm have been assigned to this new system.

These bands have been classified into two subsystem arising from the two components of $3\prod_g$ i.e., $\Omega = 0_g^\pm, 2_g$ while the spectrum corresponding to third component of $^3\prod_g$ i.e., $\Omega = 1_g$ could not be observed in this region. Bender et al [7] have been theoretically predicted that the difference of these two components may be about 300 cm^{-1} while in the present studies this separation has been found to be 359 cm^{-1} in ν_{00} values and very close to the predicted value.

3.4.   Spectrum of $F \to 2_g(^3\prod_g)$ Transition

Sixteen red degraded bands have been assigned for this transition. These bands have been classified as $\Delta_v = 0, \pm1, \pm2, \pm3, \pm4, \pm5, \pm6, \pm7, \pm8, \pm9, \pm10, \pm11$ with $(0,0)$ at 409.52nm. The vibrational constants for lower state are same as in system $E \to A$ and very close to that suggested Kedzereski et al [9-15]. The relative intensity of $F \to 2_g$ transition is less as compared to $F \to 0_g^\pm$ because of much smaller moment for $F \to 2_g$ as compared to $F \to 0_g^\pm$ transition. The molecular constants for this subsystem are as follows.

$$\nu_{00} = 24770.0 \text{cm}^{-1}, \quad \omega_e' = 151.6, \quad \omega_e' X_e' = 1.50, \quad \omega_e'' = 222.0, \quad \omega_e'' X_e'' = 1.50$$

Spectrum for this transition has been shown in Figure 4.

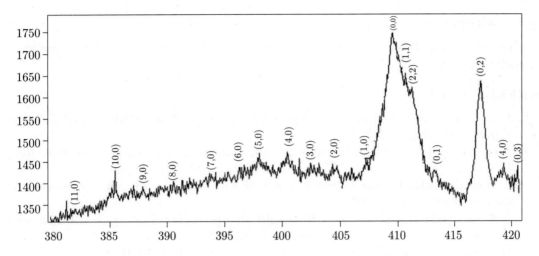

Figure 4.

3.5. Spectrum of $F \rightarrow 0_g^\pm (^3\Pi_g)$ Transition:

These bands have been assigned into $\Delta v = 0, 1, -2, -3$ and -4 sequences with $(0,0)$ lying at 403.37nm. Sixteen red degraded bands have been identified for this subsystem. The vibrational constants determined for the upper and lower state of this subsystem are same as for F2g and given as follows.

$$\nu_{00} = 24770.0 \text{cm}^{-1}, \quad \omega_e' = 151.6, \quad \omega_e' x_e' = 1.50, \quad \omega_e'' 222.0, \quad \omega_e'' X_e'' = 1.50$$

The spectrum of the system has been presented in Figure 5.

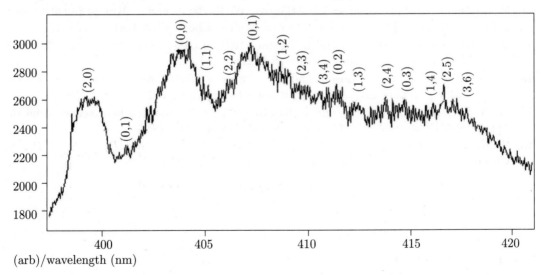

(arb)/wavelength (nm)

Figure 5. Laser produced spectrum of $F \rightarrow 0_g^\pm (^3\Pi_g)$ transition

Acknowledgements

Authors are thankful to Department of Science and Technology, New Delhi for providing financial support.

REFERENCES

1. J. G. Winans (1929), Flutings in Absorption Spectra of a mixture of Mercury and Cadmium Vapors, Phil. Mag., 7, 202.

2. J. G. Winans (1931), Properties of some zinc, cadmium and mercury bands, Phy. Rev., 37, 902.

3. H. Hamada (1931), On the molecular spectra of mercury, zinc, cadmium, magnesium and thallium, Phil. Mag., 12, 50.

4. B. S. Ault and L. Andrews (1977), Ultraviolet absorption spectra of Zn_2 and Cd_2 in solid argon and krypton at 100k, J. Mol Spectrosc., 65, 102.

5. M. Czajkowski, R. Bobkowski and L. Krause (1990), Phy. Rev. A, 41, 277.

6. P. H. Jeffery and H. D. Thom (1976), Electronic States of Zn_2, Ab initio calculations of a protptype for Hg_2, Journal of Chemical Physics, 65, 2679.

7. C. F. Bender, T. N. Rescigno, H. F. Shaefer and A.E. Orel (1979), Potential energy curves for diatomic zinc and cadmium, J. Chem. Phys., 71, 1122.

8. M. Couty, G. Chambaud and W.E. Baylis (1990), Abstract of the 12th International Conference on Atomic Physics, University of Michigan, Ann Arbor, USA; Sienkiewcz J. and Baylis W. E. (1991), private communication.

9. W. Kedzierski, J. B. Atkinson and L. Krause (1989), Laser-induced fluorescence of Zn_2 Excimer, Optics Letters 14, 607.

10. W. Kedzierski, J. Supronowicz, J. B. Atkinson and L. Krause (1990), Laser Spectroscopy of the $\Sigma_u^+(4^1P)$ states in Zn_2, Chemical Physics Letters, 173, 282.

11. W. Kedzierski, J. Supronowicz, J.B Atkinson, W. E. Baylis and L. Krause (1990), Laser-induced $(^1\Sigma_u^+) \leftarrow {}^3\prod_g$ excitation spectrum of Zn_2, Chem. Phys Letters, 175, 221.

12. W. Kedzierski, J. B. Atkinson and L. Krause (1991), Laser-induced fluorescence from the $^3\prod_u(4^3D)$ state of Zn_2, Chemical Physics Letters, 181, 427.

13. W. Kedzierski, J. B. Atkinson and L. Krause (1992), Laser spectroscopy of the $^3\Sigma_u^+(4^3P,4^3P)$ state of Zn_2, Chemical Physics Letters, 200, 103.

14. W. Kedzierski, J. B. Atkinson and L. Krause (1993) Laser-induced fluorescence from the $^3\prod_u(4^3P,4^3P)$ state of Zn_2, Chemical Physics Letters, 215, 185.

15. W. Kedzierski, J. B. Atkinson and L. Krause (1994), The $^3\Sigma_g^+(4^3P,4^3P) \leftrightarrow {}^3\Sigma_g^+(4^3P,4^1S)$ vibronic spectrum of Zn_2, Chemical Physics Letters, 222, 146.

Atoms and Molecules in Laser and External Fields

Editor: Man Mohan

Nonlinear Optics of Ultra-Short Laser Pulses: Cascades and Sliding Resonances

J. T. Mendonça

CFP, Instituto Superior Tcnico, 1049-001 Lisboa, Portugal

1. INTRODUCTION

In Nonlinear Optics of ultra-short laser pulses, two different kinds of processes can be considered: (i) the wave mixing processes, such as parametric decay, harmonic generation and three or four wave mixing [1, 2]; (ii) phase modulation processes, such as self, cross or mixed phase modulation [3], which can also be described in terms of photon acceleration [8]. This second type of processes is only relevant for short laser pulses and cannot take place for cw laser beams.

The first type of processes obeys well defined quantum relations, or matching conditions, for the frequencies and the wave-vectors of the interacting field modes. As a consequence, the frequency spectrum resulting from these processes is characterized by well defined peaks of radiation. One of the most spectacular examples of a resonant process is the occurrence of cascades of four-wave mixing, such as those observed by Crespo et al [5]. The theoretical models that can explain the occurrence of such nonlinear cascades are based on the wave equation where third or forth order polarizability terms are included [6], as described in section 2. High harmonic generation can also be included in the same model, as a particular case of these nonlinear cascades. Recent experimental work [7] shows results that are in very good agreement with our theoretical predictions.

In contrast with nonlinear wave mixing, the second type of processes to be considered here is not determined by any obvious quantum relation, and the resulting spectra depend mainly on the interaction time and on the laser pulse duration. They can be associated with an intensity dependent refractive index, and lead to the emission of a broadband specrum, sometimes called a supercontinuum. Because of their non-resonant properties, they can easily be described by photon kinetic equations [8], based on an improved version of the geometric optics approximation. Following this wave kinetics approach, we are able to derive the relevant features of the so-called self-phase modulation, without taking the wave phase into account. For this reason, we prefer to describe the effects currently known as phase modulation in terms of photon acceleration. Other related phenomena are cross phase modulation and time refraction [10], which can also be described by the photon acceleration concept [4]. Our theoretical models will be briefly discussed in section 3.

In general terms we can say that the nonlinear optics of laser pulses is dominated by resonant wave mixing processes for long pulses. In contrast, phase modulation or photon acceleration will dominate for short or ultra-short laser pulses. To these two types of processes, a third type of hybrid processes can also be added, which involves the simultaneous action of the previous two. In particular, the non-resonant spectral changes induced by photon acceleration (or phase modulation) can eventually provide the necessary matching condition for the occurrence of nonlinear wave mixing that would not take place otherwise. Because, in such case, wave mixing will be dynamically achieved in a limited area of the nonlinear medium, we have called it a sliding resonance [9]. It corresponds to nonlinear mixing that can only occur if the spectral frequency shifts provided by (self or cross) phase modulation, or more generally by photon acceleration in a non-stationary optical medium, lead to exact or nearly exact matching conditions for a significant part of the radiation spectrum. It can be shown that these sliding resonances lead to nonlinear phase-locking. The theoretical model of sliding resonances, summarized in section 4, is based on a nonlinear wave equation and gives a simple and plausible explanation for the nonlinear phase-looking observed in recent experiments.

2. NONLINEAR WAVE CASCADES

Let us start with the propagation equation for the transverse electric field E in a nonlinear optical medium. We will assume that a two laser pulses, with frequencies ω_0 and ω_1 are made to interact in the medium, and that a non-degenerate cascade of wave mixing processes takes place, as suggested by the experiments [5]. The total electric field will then be adequately described by

$$\vec{E}(\vec{r}, t) = \vec{E}\left(\sum_j\right)^N \exp(i\vec{k}\vec{r} - i\omega_j t) \tag{1}$$

with $N \gg 1$ and frequencies determined by

$$\omega_j = \omega_0 + j(\omega_1 - \omega_0) \tag{2}$$

An envelope equation for the different field components can then be established, in the form

$$2i\left[\vec{k}j \cdot \nabla + \frac{\omega_j}{c^2}n^2(\omega_j)\frac{\partial}{\partial t}\right]\vec{E} = -\frac{\omega_j^2}{\varepsilon_o c^2}\vec{P}_{NL} = (\omega_j) \tag{3}$$

where the nonlinear polarization can be written as

$$\vec{P}_{NL}(\omega_j) = \varepsilon_0(\chi\vec{E}_0^* \cdot \vec{E}_1\vec{E}_{j-1}e^{i\Delta k^+ \cdot r} + \chi'\vec{E}_0 \cdot \vec{E}_1^*\vec{E}_{j+1}e^{i\Delta k^- r}) \tag{4}$$

where χ and χ' are third order nonlinear susceptibilities. In this expression we have assumed that the field amplitudes of the two laser pulses, E_0 and E_1 are dominant over the other field components E_j, for $\omega_j \neq \omega_0\omega_1$ The phase mismatching factors are defined as

$$\Delta \vec{k}^{\pm} = \mp \vec{k}_0 \pm \vec{k}_1 + \vec{k}_{j \mp 1} - \vec{k}_j \qquad (5)$$

In the one-dimensional case, the coupled system of equations (3)-(4) can be simplified and leads to [6]

$$\frac{d}{d\tau} E_j = iw(e^{i\delta} E_{j-1} + e^{-i\delta} E_{j+1}) \qquad (6)$$

where t is a new time variable, d is a phase factor, and the coupling constant w is proportional to the square rooth of the intensities of the dominant modes E_0 and E_1. For a large number of interacting modes $N \gg 1$, this can be solved in terms of Bessel functions, according to expression

$$E_j(z) = i^{|j|} e^{ij\delta} E_0 J_{|j|}(z) - i^{(|j-1|)\delta} E_1 J_{|j-1|}(z) \qquad (7)$$

where $z = (8/3c)\omega_n n_0^2 n_2 \sqrt{I_o} I_1$. From this simple solution we can derive the expected spectrum for the nonlinear cascade, which agrees remarkably well with the observations, even when the propagation is not exactly one-dimensional, as represented in Figures 1 and 2. The cascade will only be interrupted when the dephasing Δk_j becomes larger than the spectral bandwidth of the interacting laser pulses. This means that, for shorter pulses will will have a larger number of cascaded beams. Experiments have shown that wave mixing of two laser pulses in the visible in a nonlinear optical medium such as common glass, with a duration less than 100 femto-seconds, is able to generate 20 secondary beams, corresponding to four down-shifted beams, $j_{min} = -4$, and 16 up-shifted ones, $j_{max} = 16$, and an energy efficiency conversion larger than 10 percent [5]. These beams extend from the infrared to the ultra-violet at conserve phase coherence, which suggest their use for future pulse compression, and good candidates for the production of single cycle laser pulses in the visible [11].

3. PHASE MODULATION

Let us now consider the non-resonant processes associated with optical phenomena of ultra-short laser pulses. In contrast with the previous case, we propose here the use of photon kinetics [8]. Our starting point will be the kinetic equation for photons, that can be derived from the full nonlinear wave equations in the case of a large spectrum and in the geometric optics approcimations [4]. This equation is nothing but the conservation equation for the number of photons, and can be written as

$$\left[\frac{\partial}{\partial t} + \vec{v}_k \cdot \nabla + \frac{dk}{dt} \cdot \frac{\partial}{\partial r} \right] N(\vec{r}, t) = 0 \qquad (8)$$

where $N_k(\vec{r}, t)$ is the photon number distribution, and $\vec{v}_k$ the group velocity corresponding to photons with wave-vector k. The corresponding frequency is given by the nonlinear dispersion relation of the medium

$$\omega_k = \frac{kc}{n} = \frac{kc}{n_0 + n_2 I(\vec{r}, t)} \qquad (9)$$

where n_0 and n_2 are the linear and the nonlinear refractive indices, and the intensity is given by aan integral over the spetrum $I(\vec{r}, t) = \int^- \hbar\omega_k(\vec{r}, t)d\vec{k}/(2pi)^3$.

At this point it should be noticed that the frequency for each photon mode wk is not a constant of motion, because of the time dependence of the intensity. Each photon trajectory will evolve according to the ray equations

$$\frac{d\vec{r}}{dt} = \frac{d\omega_k}{\partial \vec{k}} = \vec{v}_k, \quad \frac{dk}{dt} = -\frac{d\omega_k}{\partial k} \tag{10}$$

It can easily be recognized that these equations coincide with the characteristics of the photon kinetic equation (8). Therefore, the photon frequency shift, or photon acceleration, predicted by these ray equations is implicit in the wave kinetic equation. Analytical solutions of this equation show that the average frequency of an optical pulse propagating along a given direction Ox in a nonlinear medium evolves according to [8].

$$\langle\omega\rangle(\eta, t) = \langle\omega\rangle_0 \exp\left[\Phi\frac{\partial I(\eta)}{\partial\eta}(t - t_0)\right] \tag{11}$$

where $\eta = x - (c/n_0)t$, $< \omega >_0$ is the initial value of the frequency distribution, $\Phi = cn_2/n_0^2$ where n_0 and n_2 are the linear and the nonlinear refractive indices of the medium, and $I(\eta)$ is the pulse intensity profile. We can see that frequency up-shifts and down-shifts are expected, according to the sign of the derivative of this intensity profile. This can be understood as resulting from forces with opposite signs acting on the individual photon trajectories, as expected for the photon acceleration process. According to the above expression, for a gaussian pulsed of width σ and maximum intensity I_0, the maximum and minimum values of the photon frequency shifts will be determined by

$$\langle\omega\rangle_{\max} = \langle\omega\rangle_0 \exp\left[\mp\frac{2\sqrt{2}}{\sigma}\Phi I_0 e^{-1/2}(t - t_0)\right] \tag{12}$$

This shows an asymmetry between up and down-shifts, because the up-shifted frequency can increase indefinitely and the down-shifted ones can only tend to zero. If we consider the approximate situation, valid for short time intervals $\Delta = t - t_0$, or to relatively low intensities I_0, we recover the well known self00 phase modulation result of frequency shifts $\Delta_\omega = \langle\omega\rangle \pm -\langle\omega\rangle_0$ proportional to the time interval, as given by

$$\Delta\omega \simeq \mp\langle\omega\rangle_0\frac{2\sqrt{2}}{\sigma}\frac{cn2}{n_0^2}I_0 e^{-1/2}\Delta\tau \tag{13}$$

This shows that the so called phase modulation effect can be recovered by using our photon kinetic approach, where the phase information was not retained. A more adequate physical picture of this process should then be based on photon acceleration, which is also able to describe cross-modulation for co and counterpropagating laser pulses, as well as frequency shifts due to other causes (linear or nonlinear) of time dependent optical forces acting on the photon trajectories, as explicitly shown by equations (10) [4].

4. SLIDING RESONANCES

The third type of nonlinear processes of ultra-short laser pulses combines the other two. It corresponds to the occurrence of nonlinear wave mixing in geometric con-figurations that would prevent them in the case of long or cw laser pulses. This occurs with ultra-short laser pulses because the phase-matching conditions, that initially are not satisfied, become possible due to the spectral changes associated with photon acceleration or phase modulation. In order to illustrate this, let us consider three optical beams, with central frequencies ω_0, ω_1 and ω_2, propagating in a nonlinear medium such that, initially, their frequency and phase matching conditions are not satisfied. They behave quite independently from each other, unless their short duration and high intensities allow for self and cross-modulation to take place. As a result, their frequency spectrum will change along propagation in such a way that, at some given position and time $t = \tau$ they can eventually become resonant and phase-matched:

$$\Delta\omega(\tau) = \omega_0(\tau) - \omega_1(\tau) - \omega_2(\tau), \Delta\vec{k}(\tau) = \vec{k}_0(\tau) - \vec{k}_1(\tau) - \vec{k}_2(\tau) \tag{14}$$

Using a new variable $\eta = z - v_0 t$ and assuming envelope solutions for the amplitudes $A_j(\eta, t) \exp[i\phi(\eta, t)]$

$$\frac{1}{v_0}\left(\frac{\partial a_0}{\partial t} + ia_0\frac{\partial\phi_0}{\partial t}\right) = i\frac{\omega_0^2}{2k_0c^2}\left[\chi^{(2)}a_1a_2e^{is(\eta,t)} + \chi^{(3)}(|a_1|^2 + |a_2|^2)a_0\right] \tag{15}$$

where the total phase mismatch $s(\eta, t)$ is given by:

$$s(\eta, \tau) = \Delta k(\eta + v_0\tau) - \Delta\omega\tau + [\phi_1(\eta, \tau) + \phi_2(\eta, \tau) - \phi_0(\eta - \tau)] \tag{16}$$

Equating to zero the terms proportional to the envelope amplitude a_0, we obtain two separate equations for the amplitude and phase of the mode $j = 0$, of the form:

$$\frac{\partial\phi_0}{\partial t} = \alpha_0\omega_0(|a_1|^2 + |a_2|^2) \tag{17}$$

and:

$$\frac{\partial a_0}{\partial t} = iC_0a_1a_2e^{is(\eta,\tau)} \tag{18}$$

with the nonlinear coefficients α_0 and C_0.

It is obvious from here that equation (17) describes the cross-phase modulation of the optical pulse $j = 0$ due to the presence of the other two $j = 1, 2$. We can also recognize that equation (18) describes the nonlinear three-wave coupling. This nonlinear coupling can only become significant if, for some range of value η and t, the total phase mismatch is reduced nearly to zero: $s(\eta, t) \simeq 0$. When that occurs, the variation of the wave amplitudes a_j will have a feedback influence on the phase modulation. The two effects, phase modulation and three-wave coupling, are therefore coupled to each other though the equations (17) and (18). Of course, these equations have to be complemented by those associated with the other two modes, which can similarly be written as

$$\frac{\partial\phi_j}{\partial t} + \left(1 - \frac{v_j}{v_0}\right)\frac{\partial\phi_j}{\partial\eta} = \alpha_j\omega_j(|a_0|^2 + |a_{k\neq j}|^2) \tag{19}$$

and, for the amplitudes:

$$\frac{\partial a_j}{\partial t} + \left(1 - \frac{v_j}{v_0}\right)\frac{\partial a_j}{\partial \eta} = iC_j a_0 a_{k\neq j}^* e^{-is(\eta,t)} \tag{20}$$

where j and k can take the values 1 and 2. The cross-phase modulation coefficients α_j, and the wave mixing coefficients C_j for $j = 0, 1, 2$, are determined by:

$$\alpha_j = \frac{\omega_j v_j}{2k_j c^2}\chi^{(3)}, \quad C_j = \frac{\omega_j^2 v_j}{2k_j c^2}\chi^{(2)} \tag{21}$$

Similar equations could also be derived for sliding resonances of four-wave mixing, which would be relevant for centro-symmetric optical media, where the second order susceptibility is not present $\chi^{(2)} = 0$.

In order to understand the physical relevance of the sliding resonance, and to illustrate its main features, we will consider the simple but important case of parametric amplification of a small optical signal a_1 by a strong wave a_0. From equations (20) we can easily get

$$\frac{\partial^2 a_j}{\partial t^2} = k^2 a_j - i\frac{\partial s}{\partial t}\frac{\partial a_j}{\partial t} \tag{22}$$

where $k^2 = C_2 C_1^* |a_0|^2$ Let us first consider the case where the phase modulation effects are negligible, which corresponds to $\alpha = 0$. We can then replace $(\partial s/\partial t) = -\Delta\omega(0)$ in equations (22). The general solution of these equations, valid for the initial conditions: $a_2(0) = 0$ and $(\partial a_1(0)/\partial t) = 0$, is:

$$a_1(t) = a_1(0)\cosh\left(\sqrt{k^2 - \Delta\omega(0)/4t}\right)\exp[i\Delta\omega(0)t/2] \tag{23}$$

and:

$$a_1(t) = ia_1(0)\sinh\left(\sqrt{k^2 - \Delta\omega(0)/4t}\right)\exp[i\Delta\omega(0)t/2] \tag{24}$$

For a perfect frequency matching $\Delta w(0) = 0$ this would reduce to the well known results for parametric amplification: $a_1(t) = a_1(0)\cosh(kt)$ and $a_2(t) = ia_1(0)\sinh(kt)$. However, for a non negligible frequency mismatch we would simply have very small oscillations of $a_1(0)$ around its initial value, with no noticeable energy transfer from the wave pump a_0. In contrast, in the presence of a phase modulation term $\alpha \neq 0$, a sliding resonance will take place after some time $t = t_0$, where the nonlinear energy transfer between the different field modes can be significant. In this case equation (22) can be written in the form of a Hermite equation

$$\frac{\partial^2 a_j}{\partial t^2} = k^2 a_j + i\alpha(2t - t_0)\frac{\partial a_j}{\partial t} \tag{25}$$

for the modes $j = 1, 2$. Its solutions can be written in terms of confluent hypergeometric functions [9].

This is illustrated in Figure 3, where the existence of a frequency mismatch of order k^2 prevents parametric decay but the inclusion of the phase modulation terms will produce a sliding resonance that will reinstall parametric amplification after some initial delay.

Similar results could also be obtained with more general three or four wave interactions. Recent experimental results, using a configuration similar to that used to produce the nonlinear optical cascades described in Section 2, but with an interaction angle that would prevent resonant wave mixing, we where able to observe the formation of similar nonlinear cascades, each of the cascaded modes showing a considerable spectral broadening, after a much longer interaction time [11]. In this case, the decisive influence of cross phase modulation was experimentally verified, in agreement with the theoretical model described in this section.

5. CONCLUSIONS

The main types of nonlinear optical processes of ultra-short laser pulses were identified, and three different theoretical models were described. More details are given in the original papers. These models can provide a very accurate description of the different kinds of experiments associated will nonlinear optical cascades, phase modulation, photon acceleration and sliding resonances. Three different theoretical models and experimental configurations were considered. In the first one, corresponding to the nonlinear cascades, a system of nonlinear coupled equations was used, and analytical solutions were obtained for simple but physically relevant situations. A different model was used to describe the frequency shifts and spectral broadening known as phase modulation or photon acceleration. In this case we have used a wave kinetic description, which is an elaborate kinetic version of geometric optics, and have shown that the field phase is not relevant for the description of the so called phase modulation. Finally, sliding resonance is an hybrid process where both the resonant and the non-resonant nonlinear terms are relevant and have to be taken into account in the nonlinear wave equations. It should also be stressed that the three types of nonlinear processes described here can preserve phase coherent, which makes them very attractive for the development of future sources of sub-femtosecond and of single cycle laser pulses centered in the visible. Two distinct experimental configuration based on these methods are currently being explored.

REFERENCES

1. R.W. Boyd, Nonlinear Optics, Academic Press, San Diego (1991).
2. K.S. Ho et al., Physics of Nonlinear Optics, World Scientific, Singapure (2000).
3. Alfano (editor), The Supercontinuum Laser Source, Springer-Verlag, New York (1989).
4. J.T. Mendonça, Theory of Photon Acceleration, Institute of Physics Publishing, Bristol (2001).
5. H. Crespo, A. Dos Santos and J.T. Mendonça, Opt. Lett., 25, 829 (2000).
6. J.T. Mendonça, A. Guerreiro and H. Crespo, Opt. Commun., 188, 383 (2001).
7. L. Misoguti et al., Phys. Rev. Lett., 87, 013601 (2001).
8. L.O. Silva and J.T. Mendonça, Opt. Commun., 196, 285 (2001).
9. J.T. Mendonça and H. Crespo, Opt. Commun., 222, 405 (2003).
10. J.T. Mendonça, A.M. Martins and A. Guerreiro, Phys. Rev. A, 68, 043801 (2003).
11. H. Crespo, Ph.D. thesis, IST, Lisbon (2006).

Atoms and Molecules in Laser and External Fields

Editor: Man Mohan

Copyright © 2008, Narosa Publishing House, New Delhi, India

Coherent Control of Population Transfer to Excited Vibrational-Rotational Levels by Stimulated Raman Processes

Chitrakshya Sarkar, Rangana Bhattacharya and Samir Saha

Atomic and Molecular Physics Section, Department of Materials Science,
Indian Association for the Cultivation of Science, Jadavpur, Kolkata 700032, India

1. INTRODUCTION

The study of efficient and selective population transfer to excited atomic or molecular levels, which are not accessible by one-photon transition, is of current interest in many applications, such as collision dynamics, spectroscopy and optical control of chemical reactions. One of the elegant methods based on stimulated Raman processes producing almost complete population transfer (or population inversion) to excited vibrational-rotational levels in diatomic molecules by pump and Stokes laser pulses is *Stimulated Raman Adiabatic Passage* (STIRAP) [1-4]. STIRAP is a stimulated resonant or near-resonant Raman technique which usually requires 3 conditions to be satisfied: (i) two-photon Raman resonance,(ii) counterintuitive pulse order with partial overlap between the pulses, and (iii) adiabatic time evolution. Through these three conditions, one can force almost complete population inversion from the initially populated ground level to a final target level of a three- or multi-level system, without transfer of any appreciable population in the intermediate level (or levels). For a 3-level Λ system (Figure 1) [2]:

$\Omega_P(t)$ and $\Omega_S(t)$ are the time-dependent one-photon coupling matrix elements for Rabi transitions by the pump (P) and Stokes (S) fields, Γ_{gi} and Γ_{fi} are the (spontaneous) decay rates from level $|i\rangle$ to levels $|g\rangle$ and $|f\rangle$, Γ_{out} is the (spontaneous or induced) decay rate out of the Λ system, E_g, E_i and E_f are the eigenenergies of the (unperturbed) molecular levels.

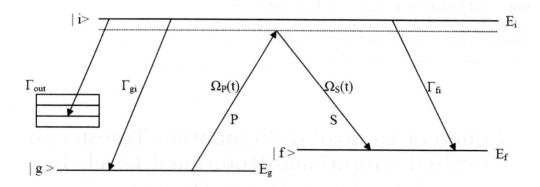

Figure 1. Energy-level diagram of a three-level system

The populations (ρ_g, ρ_i, ρ_f) of the levels are obtained by solving the Liouville equation of motion of the density matrix operator $\rho(t)$ given by (in a.u.) [4]

$$d\rho/dt = -i[H(t), \rho(t)]\Gamma\rho(t). \tag{1}$$

where $H(t) = H_0 + H_1(t)$ is the total Hamiltonian of the system (molecule + laser fields), $H_0 = H_M + H_\Gamma$ is the Hamiltonian of the (unperturbed) molecule + fields, $H_1(t) = -E(t)d$ (in the electric field or length gauge form with dipole approx.) is the interaction Hamiltonian between the molecule and the fields, d is the transition electric dipole moment operator whereas $E(t) = E_P(t) + E_S(t)$ is the total electric field vector. The classical forms of the pump and Stokes fields are

$$E_P(t) = (1/2)E_P^0 f_P(t)[\varepsilon P \exp(-i\omega_P t) + \varepsilon P^* \exp(i\omega_P t)] \tag{2}$$

$$E_S(t) = (1/2)E_S^0 f_S(t)[\varepsilon S \exp(-i\omega_S t) + \varepsilon S^* \exp(i\omega_S t)], \tag{3}$$

where ω_P and ω_S are the frequencies of the (assumed) single mode laser fields, E_P^0 and E_S^0 are the peak amplitudes of the fields, ε_P and ε_S are the unit polarization vectors of the fields whereas $f_P(t)$ and $f_S(t)$ are the envelopes (or profiles) of the field amplitudes. For Gaussian pulses:

$$f_P(t) = \exp[-t^2/2\sigma P^2] \tag{4}$$

and

$$f_S(t) = \exp[-(t - \Delta t)^2/2\sigma S^2], \tag{5}$$

where σ_P and σ_S are the pulse widths (FWHM) and Δ_t is the time delay between the pulses (Figure 2).

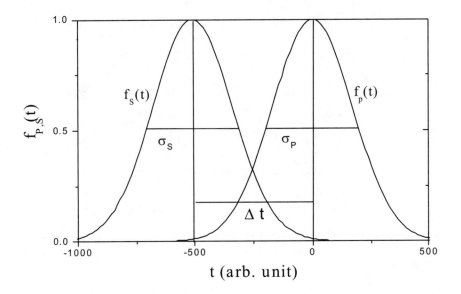

Figure 2. Variation of the pump and Stokes field profiles ($f_P(t)$, $f_S(t)$) with pulse time (t)

For counterintuitive pulse order, $\Delta t = -ve$ whereas for intuitive pulse order, $\Delta t = +ve$ For adiabatic time evolution [1],

$$\Omega^0_{eff}\Delta t \gg 1, \tag{6}$$

where

$$\Omega^0_{eff} = [(\Omega^0_P)^2 + (\Omega^0_S)^2]^{1/2} \tag{7}$$

at peak amplitudes (E^0_P, E^0_S). The Rabi frequencies are

$$\Omega_P(t) = (1/2)\langle i|E^0_P f_P(t)(\varepsilon_P + \varepsilon^*_P)d|g\rangle \tag{8}$$

and

$$\Omega_S(t) = (1/2)\langle f|E^0_S f S(t)(\varepsilon_S + \varepsilon^*_S)d|i\rangle \tag{9}$$

For two-photon Raman resonance,

$$\delta = (E_f - E_g) - (\omega_P - w_S) = 0. \tag{10}$$

Ω_{ji} denotes the time-dependent one-photon coupling matrix element for Rabi transitions from level $|i\rangle$ to level $|j\rangle$ by the pump or Stokes field. V_{12} is the nonadiabatic interaction matrix element between states $|1\rangle$ and $|2\rangle$. $\gamma_{01}(\gamma_{02})$ and $\gamma_{31}(\gamma_{32})$ denote the spontaneous radiative decay rates from level $|1\rangle$ ($|2\rangle$) to levels $|0\rangle$ and $|3\rangle$, γ_{out1} (γ_{out2}) denotes the decay rates of level $|1\rangle$ ($|2\rangle$) to other levels out of the four-level system considered.

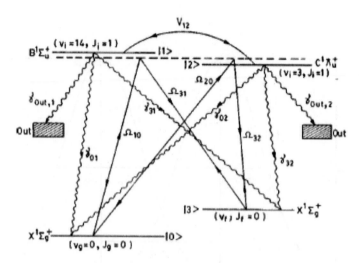

Figure 3. Schematic diagram of the four-level H_2 system interacting with the pump and Stokes lasers where all state levels, the relevant Rabi frequencies, as well as the spontaneous decay rates are indicated.[1]

Numerical solutions of the density matrix equations [4] give populations (P) of the initial (g), intermediate (i) and final (f) levels (Figure 4).

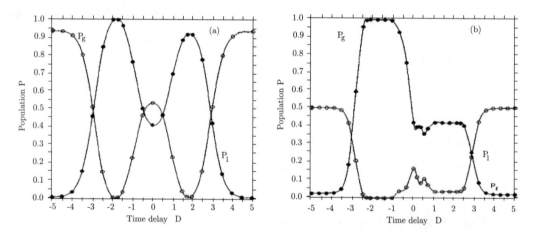

Figure 4. Populations of the ground (P_g) and final (P_f) levels (after the pump and Stokes laser pulses are over) plotted against (reduced) time delay ($D = \Delta t/\sigma_P$) between the pulses for the Q-branch ($J_r = 0$) fundamental ($v_f = 1$) transition from $X(v_g = 0, J_g = 0)$ level for resonance with the intermediate perturbed $B(14, 1)$ level for $\sigma_P(\sigma_S) = 170ns$ and different values of $I_P^0(I_S^0)$ including the NA interaction (V_{12}) between $B(14,1)$ and $C(3,1)$ levels. (a)$I_P^0(I_S^0) = 1 \times 10^6$ W/cm^2, (b)$I_P^0(I_S^0) = 1 \times 10^7$ W/cm^2 with λ_P and λ_S and 95 nm.[1]

---

[1]Reprinted with permission from S.Ghosh, S. Sen, S.S. Bhattacharyya, ans S. Saha, Phys. Rev. A 59, 4475 (1999), 1999, APS.

An alternative technique for population transfer is Chirped Adiabatic Raman Passage (CARP) [5-7]. CARP is a stimulated nonresonant Raman process. In this method: (i) the pump and Stokes laser pulses $(E_P(t), E_S(t))$ are applied simultaneously (without any time delay between the pulses unlike STIRAP),(ii) the pump and Stokes laser frequencies (ω_P, ω_S) are far from one-photon resonance (unlike STIRAP), (iii) the laser frequencies are chirped (vary with time unlike STIRAP), (iv) two-photon Raman resonance is not always to be satisfied throughout the pulse periods (unlike STIRAP), and (v) adiabatic time evolution follows (like STIRAP). Thus CARP and STIRAP are of opposite nature (in some sense) but goal is the same (to produce complete population inversion). In CARP, the laser frequencies are chirped linearly as [7]

$$\omega_P(t) = \omega_P \pm c_r^P t \tag{11}$$

and

$$\omega_S(t) = \omega_S \mp \Delta\omega_S \pm c_r^S t, \tag{12}$$

where $\Delta\omega_P(\Delta\omega_S)$ are the initial $(t=0)$ sweeping $= \gamma\Omega^{0(2)}$ at the peak field amplitudes (E_P^0, E_S^0), c_r^P (c_r^S) are chirp rates $= \beta(\Omega^{0(2)})^2, \gamma$ is the dimensionless chirp width parameter $(>1), \beta$ is the dimensionless chirp rate parameter (<1) and $\Omega^{(2)}$ is the two-photon Rabi frequency.

For H_2 system (Figure 5) [7]:

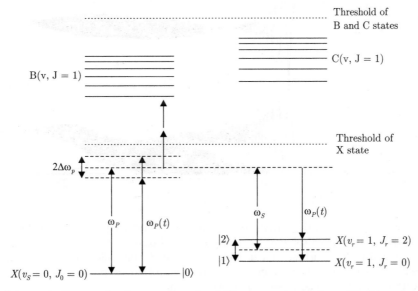

Figure 5. Schematic diagram for Raman three-level H_2 system intracting with chirped pump $[\omega_P(t)]$ and Stokes $[\omega_S(t)]$ laser pulses where all state levels and relevant transitions are specified.[2]

The pump and Stokes lasers of central frequencies ω_P and ω_S are far from resonance with the upper nonresonant states with a minimum detuning Δ. $2\Delta\omega_p$ $(2\Delta\omega_S)$ is the total chirping of the pump (Stokes) pulse.

Dynamic Stark shifts [7] of the initial and final levels due to two-photon Raman transitions via nonresonant intermediate states are incorporated in the Liouville equation of the density matrix operator $\rho(t)$. Numerical solutions of the density matrix equations [7] give the populations (Figure 6) of the initial and final levels.

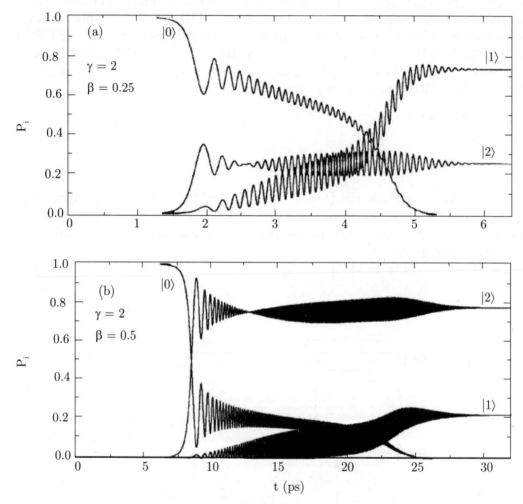

Figure 6. Populations (P_i) of the initial ($|0>$) and final ($|1>$ and $|2>$) levels as functions of time $t(ps)$ for negative chirping of the pump laser and positive chirping of the Stokes laser with γ (dimensionless chirp width parameter) $= 2$ and β (dimensionless chirp rate parameter) $= 0.25$ (a), 0.05 (b). The peak intensity of the pump (Stokes) pulse is $I_P^0(I_S^0) = 100$ TW/cm^2 with central wave-length $\lambda_P(\lambda_S) = 532(691)$ nm.[2]

(Raman active) small molecules require relatively high-energy (UV/VUV) pump and Stokes photons to reach the first excited (resonant intermediate) electronic states

---

[2]Reprinted with permission from S. Sen, S. Ghosh, S.S. Bhattacharyya and S. Saha, J. Chem. Phys. 116, 581 (2002), 2002, AIP.

[4]. It is generally difficult to provide UV/VUV lasers with adequate power and pulse duration for the application of the $(1+1)$-photon STIRAP process. A natural extension of STIRAP is STIHRAP (Stimulated Hyper-Raman Adiabatic Passage) [8] where the usual Raman transitions in STIRAP are replaced by hyper-Raman transitions.

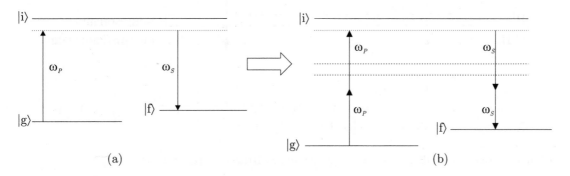

Figure 7. Energylevel diagram of a $(2+2)$-photon STIHRAP system (b) as an extension of a $(1+1)$-photon STIRAP system (a)

In STIHRAP, the initial ground (g) and the final (f) levels are very close to two-photon resonance with the resonant intermediate (i) level by pump and Stokes laser frequencies (ω_P, ω_S). ω_P and ω_S are linearly chirped as

$$\omega_P(t) = \omega_P + c_r^P t \tag{13}$$

and

$$\omega_S(t) = \omega_S + c_r^S t, \tag{14}$$

where c_r^P is the chirp rate of the pump frequency $(\omega_P) = \beta_P(S_i^0 - S_g^0)/\sigma_P$ and c_r^S is the chirp rate of the Stokes frequency $(\omega_S) = \beta_S(S_i^0 S_f^0)/\sigma_S$, S_g^0, S_i^0 and S_f^0 are the dynamic Stark shifts (due to two-photon transitions via intermediate nonresonant states) of the ground (g), (resonant) intermediate (i) and final (f) levels at the peak field amplitudes (E_P^0, E_S^0), σ_P and σ_S are the pump and Stokes pulse widths (FWHM), and β_P, β_S are the dimensionless chirp rate parameters (< 1). In the STIHRAP process, (i) the pump and Stokes pulse are applied simultaneously (unlike STIRAP but like CARP), (ii) two-step Raman resonance is not always satisfied throughout the pulse periods (unlike STIRAP but like CARP), and (iii) adiabatic time evolution follows (like STIRAP and CARP).

The pump and Stokes laser frequencies ω_P and ω_S are away from resonance with the intermediate nonresonant states.

The dynamic Stark shifts (S_g, S_i, S_f) of the levels and the spontaneous decay rate (Γ_i) of the intermediate resonant level are incorporated in the density matrix equations. The numerical solutions of the Liouville equation of the density operator $\rho(t)$ give the populations (Figure 9) of the ground, (resonant) intermediate and final levels.

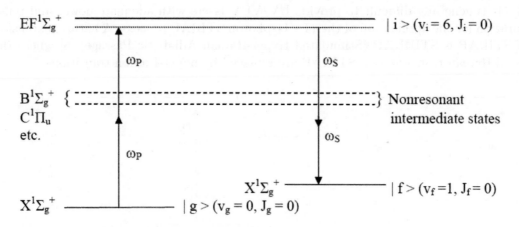

Figure 8. Energylevel diagram of a $(2+2)$-photon STIHRAP in H_2 system

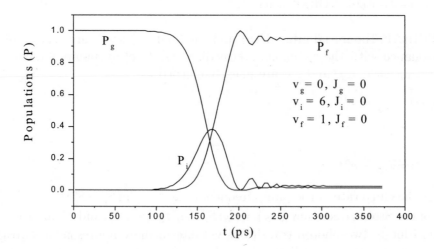

Figure 9. Populations (P) of the ground (g), (resonant) intermediate (i) and final (f) levels plotted against pulse time (t) with. The peak intensity of the pump pulse is $I_P^0 = 1.4 \times 10^{11}$ W/cm^2 and Stokes pulse is $I_S^0 = 1.0 \times 10^{12}$ W/cm^2 with initial $(t=0)$ pump (Stokes) pulse wavelength $\lambda_P(\lambda_S) = 193$ (201) nm. The dimensionless chirp rate parameters $\beta_P(\beta_S) = 0.52$

Acknowledgements

The contribution of Prof. S.S. Bhattacharyya to this work is gratefully acknowledged.

REFERENCES

1. J. K. Bergmann, H. Theuer and B.W. Shore, Rev. Mod. Phys. **70**, 1003 (1998); N.V. Vitanov, T. Halfmann, B.W. Shore and K. Bergmann, Annu. Rev. Phys. Chem. **52**, 763 (2001) and references cited therein.

2. Y.B. Band and P.S. Julienne, J. Chem. Phys. 94, 5291 (1991); 95, 5681 (1991); 97, 9107 (1992); Y.B. Band, Phys. Rev. A 45, 6643 (1992).

3. N.V. Vitanov and S. Stenholm, Opt. Commun. 127, 215 (1996); 135, 394 (1997); Phys. Rev. A 55, 648, 2982 (1997); 56, 741,1463 (1997).

4. S. Ghosh, S. Sen, S.S. Bhattacharyya and S. Saha, Phys. Rev. A 59, 4475 (1999); Pramana J. Phys. 54, 827 (2000).

5. S. Chelkowski and G.N. Gibson, Phys. Rev. A 52, R3417 (1995); S. Chelkowski and A.D. Brandrauk, Raman Spectrosc. 28, 459 (1997); F. Légaré, S. Chelkowski and A.D. Brandrauk, J. Raman Spectrosc. 31, 15 (2000).

6. J.C. Davis and W.S. Waren, J. Chem. Phys. 110, 4229 (1999) and references cited therein.

7. S. Sen, S. Ghosh, S.S. Bhattacharyya and S. Saha, J. Chem. Phys. 116, 581 (2002).

8. L.P. Yatsenko, S. Guérin, T. Halfmann, K. Böhmer, B.W. Shore and K. Bergmann, Phys. Rev. A 58, 4683 (1998); S. Guérin, L.P. Yatsenko, T. Halfmann, B.W. Shore and K. Bergmann, Phys. Rev. A 56, 4691 (1998); K. Böhmer, T. Halfmann, L.P. Yatsenko, B.W. Shore, and K. Bergmann, Phys. Rev. A 64, 023404 (2001).

9. C. Sarkar, R. Bhattacharya, S.S. Bhattacharyya and S. Saha, to be published.

Index